Women Behind the Wheel

Women Behind _the_ Wheel

An Unexpected and Personal History of the Car

Nancy A. Nichols

PEGASUS BOOKS

NEW YORK LONDON

WOMEN BEHIND THE WHEEL

Pegasus Books, Ltd.
148 West 37th Street, 13th Floor
New York, NY 10018

Copyright © 2024 by Nancy A. Nichols

First Pegasus Books cloth edition March 2024

Interior design by Maria Fernandez

Library of Congress Cataloging-in-Publication Data is available.

ISBN: 978-1-63936-559-3

10 9 8 7 6 5 4 3 2 1

Printed in the United States of America
Distributed by Simon & Schuster
www.pegasusbooks.com

For Jacob

Contents

Car culture, briefly defined, is the cluster of beliefs, attitudes, symbols, values, behavior and institutions which have grown up around the manufacture and use of automobiles. Its economic base is an enormous, many-faceted industry which leads the business cycle and has profound implications for domestic and foreign policy. It has its own subcultures which specialize in customized vans, sports cars, and the like. As an "American way of life," it invests a machine with values transcending in importance that of efficient, economic transportation. It fosters, for instance, a neo-frontier spirit, evident in the many popular films which feature cars in exciting chases, bouts with the law, daring escapes and escapades. It has its own rituals, taboos, folk songs, and legendary heroes. . . . Exemplary cars mark rites of courtship, marriage, death, and great public occasions; and romance has never been quite the same since the advent of America's "love affair" with cars. Toy cars and car games wait in the wings for children, children grow up to buy real cars—"most delicious of adult toys"—and exchange Halloween for the annual showroom display of next year's models to celebrate the harvest season. The high priest of this car culture, where cars sometimes seem to breed cars, is the American Adam. But where is Eve?

—Charles L. Sanford, "'Woman's Place' in American Car Culture"

ONE

Birth of a Car Salesman

In 1888 in Mannheim, Germany, when thirty-nine-year-old Bertha Benz grew tired of her husband's timidity and procrastination and even more tired of footing the bill for his innovative yet time-consuming invention of the automobile, she took his "Patent Motor Wagon" without his knowledge and set out at dawn to visit her mother just over sixty miles away. In that early primitive vehicle—which to modern eyes resembles little more than a giant motorized tricycle—she and her two sons took what is now widely regarded as the first ever road trip. During the twelve-hour ride, she garnered a great deal of attention for her husband's invention and surprised everyone with her mechanical know-how by using her hatpin to unclog a blocked fuel line and, in what one auto enthusiast called the "height of eroticism," sheathing the car's worn ignition cable with her garter. [1]

Yet by the time I was growing up in the Midwest in the 1960s and 1970s, that little adventurous piece of women's history was buried deep in the history books. In Waukegan, the small town on the Illinois-Wisconsin border where I was born, we just assumed that our very own Henry Ford had invented the car. At that time Detroit was not just the center of our world or the car world, it was the center of America. We pledged our allegiance to the flag each day in school, but at home we swore to that age old mid-western oath—that whatever was good for General Motors was good for America. We knew all too well that there was a direct line between the sales on the lot, the food on our dinner table, and the roof over our heads.

The garages and factories near my hometown and the massive factories in Detroit were populated by what one historian has called "men of burly practicality."[2] Men who could brave the fear of fire in the furnaces, with-stand the nauseating smell of gasoline in summer, and endure the unheated garages and snowy lots of winter. Women drivers, far from being portrayed as adventurous pioneers like Bertha Benz, were for the most part ridiculed on late night television, where jokes about their driving skills were as ubiquitous as they were around the dinner table or at the bar down at the American Legion Hall.

In the 1970s, when I was in junior high and high school, the girls at my school were explicitly banned from shop class, where auto repair was a first step toward an engineering degree, a steady job at the dealership, or a position as a machinist in the factory. Instead, we took required cooking classes, where besides mastering a meringue we had to know how to spell *colander* (one *l*, not two).

My father sold used and new cars in our town. He dressed the part of a car salesman as well as he played it. At nearly six foot four, he was unmis-takable in his lime-green leisure suits with white belts and matching shoes. His slicked-back silver hair was a perfect match for black shirts paired with white ties and checkerboard sports jackets.

He often drove a Dodge Charger, one of the most powerful vehicles ever to hit the city streets in the United States. It was a lightweight car with a big engine and loud muffler, a vehicle popular with young men just back from Vietnam. I can picture him alone in the small shack at the back of the lot, his feet perched on an old aluminum desk. Cigarette hanging from his mouth, he was slow to get up but a fast talker once he reached you.

Sometimes he worked in the dealer's new-car showroom on the other side of town. There, he sold Chryslers and Dodges under bright fluorescent lights that reflected off the showroom cars like a disco ball turned upside-down.

In 1970, when I was in the sixth grade, he sold more Dodge Darts than any other man in the state of Illinois. The company gave him a small diamond pin to mark this achievement, and he wore it religiously. The automobile industry had embraced the "new and improved" sales strategy, and each year's version of the Dart was slightly more alluring than the last. The cars all looked pretty much the same to me—the Dart was a boxy economy car that I remember in mostly pale pastel colors with shiny vinyl upholstery—but my father excelled at extolling the small virtues and changes in each new model.

Though he mainly sold used cars, he would often drive new ones from the showroom floor, tooling around in the latest models, trying to gin up interest from the factory workers in town. This meant we had a new car every week. We rode around in one stylish model after another, the price tag stuck on the rear window, small paper squares beneath our feet to protect the carpet, that sweet new car smell filling our nostrils. Each new car filled us with hope and aspiration. Our bank account may have been empty, but the gas tank was always full.

The men in my family went fast, made a lot of noise, and blew a lot of smoke in powerful gas-guzzling behemoths. Often sponsored by my father, my brother drove race cars on the weekend surrounded by women clad in halter tops and bell bottoms. He laughed when he smashed up his car, and he whooped when he beat someone to the finish line. The cars my father and brother drove were fast and loud as they belched and smoked their way out of the driveway or down the racetrack—gears clicking in a hypnotic rhythm, surging as they gained speed.

In sharp contrast, the cars my mother and sister drove were full of style—a sporty red Mustang for my sister and a Chevy convertible with tail fins for my mom. They were fashion statements made of steel.

This is a book about growing up female in the Midwest during the heyday of American car culture. It's a book about what my family drove back then and what I drive now. It's a story about how the car came to be our most gendered technology and how it both emboldened and upended women's lives in myriad uncounted and unrecognized ways. It's about what we failed to pay attention to then and what it costs us all now in terms of personal freedom, access to nature, and global environmental degradation. The story begins with my father: the car salesman.

❊

When my father was about ten, his brother, Donny, was killed in a car accident. That my father was implicated in some way in the death of his six-year-old brother seems undeniable. What exactly happened is like everything else in my family, a miasma, an impenetrable series of events from which all sorts of other crazy events unspooled. I would eventually come to understand what happened many years later, but as a child my father's story about that day changed constantly.

One version of the story goes like this: Two boys were playing pony. There was a rope around the waist of the smaller boy, Donny, who was playing the part of the pony. The older boy, my father, was pretending to ride him. The smaller boy broke free and ran into the street, where he was hit and killed by a school bus.

At other times my father would tell a different story. My father hit a ball into the street; Donny ran after it. The driver tried to brake but couldn't. Or he didn't brake at all. Maybe my dad told Donny not to go into the street, but Donny did anyhow. Maybe my father didn't say anything at all.

When my father first told me the story, it was a school bus. Later, it was an ice truck. But in the first instance, when he first told it, it was a school bus. Or perhaps it was an ice truck. But then again it might have been two boys playing pony.

And so, it began. Both blamed and punished at the time for his brother's death, my father's lying most likely began as the understandable defensive strategy of a traumatized child. Even now I can feel empathy for that small boy who told that first lie. Funerals were arranged from the home back then, and my father was made to sleep in the same room as his dead brother. He would talk about that night often even as he dissembled the events of the day.

Memories and the deep guilt and trauma of Donny's death—whatever the details, whatever the circumstances that might be construed as fact, whatever blame or guilt, real or imagined, that might have been assigned to my father—stayed with him forever. Decades of dysfunction followed.

As a grown man, my father lied about everything, consistently, reflexively, whether his lies served a purpose or not. He lied about which grocery store he went to and whether the car was insured or whether there was oil in the burner. Eventually, my father would lose all track of the truth, and with it he would lose the thread of his own life story. When my father became a car salesman, lying became his business.

❅

Selling as a profession, divorced from production or artistry or husbandry, did not emerge full force until the 1900s, when the great engine of mass production began to separate the creation of goods from their dissemination into the marketplace. The salesman as a distinct job and profession emerged just as the United States made the shift from a mostly agrarian and rural society, where we met our own needs and bartered for what we could not make ourselves, to the era of mass production, which began with the automobile itself. Indeed, the salesman played an important role in creating our vast consumer society with its endless temptations, variations, and choices.

From the beginning, however, the salesman and the role he plays has been suspect. Salesmanship, with its element of confrontation, and the ever-present threat of being swindled, is an inherently uncomfortable process. "There is something unsettling about the salesperson's telephone call for an appointment, approach on the car lot, or knock on the office door," writes business historian Walter Friedman in his book on the history of sales, *Birth of a Salesman: The Transformation of Selling in America.*

Early peddlers, Friedman writes, were portrayed in literature as a threat to farmer's daughters everywhere and were given names such as Sam Slick to reference their suspect tactics. In 1857 Herman Melville introduced a satanic riverboat traveler who sold herbal medicines in his well-known novel *The Confidence Man.* Some fifty years later, in his classic 1922 satire, *Babbitt,* Sinclair Lewis portrayed a prototypical real estate salesman, George F. Babbitt, with disdain. Babbitt, Lewis writes, "Made nothing in particular, neither butter nor shoes nor poetry, but he was nimble in the calling of selling homes for more than people could afford."[3]

But the salesman—even if he was widely disliked—also served a clear purpose. Friedman refers to salesmanship as part of a generalized search for economic order and cites an early doctoral dissertation on the subject that referred to selling as "an expansion of meaning."

Salesmen, Friedman notes, helped give meaning to our purchases and were key to the acceptance of innovative products such as the vacuum cleaner, sewing machines, shelf clocks, and packaged items such as ketchup and toothpaste. As a result, salesmen were powerful engines of economic progress, shifting both our view of the world and our sense of what we needed to live in it. "Salesmen and saleswomen are at the center of the story of industrialization, innovation, and change."[4]

To bolster his argument, Friedman cites a *Fortune* magazine article that says without the vast power of salesmen, "mass production would be a shadow of what it is today."[5] The emerging job of the salesman, therefore, was not just to move goods; the job of the salesman was to give meaning to the goods he sold. As a result, salesmen sold not only a new vision of home and hearth—what we now call lifestyles—they sold the act of consumption itself.

And they did it with fervent religious zeal. Remarkably, a 1925 bestseller, *The Man Nobody Knows*, portrayed Jesus Christ as a successful advertising and marketing executive and credited him with being no less than the founder of modern salesmanship. "Surely no one will consider us lacking in reverence if we say that every one of the principles of modern salesmanship on which businessmen so much pride themselves," wrote author Bruce Barton, "are brilliantly exemplified in Jesus' talk and work."[6]

Yet despite the obvious links with religious proselytizing, the salesman in literature was rarely seen as a positive force; early traveling salesmen were seen as sexual threats, and women only begrudgingly answered their knock on the door. That hesitancy and disdain—the functional friction between male salesman and female buyers—would last well into the era of mass commerce.

If the salesman in general has always been suspect, a figure to poke fun at or one to avoid at all costs, the car salesman, in particular, has always occupied a perilously low rung on the already low-slung ladder of salesmen. A 2020 Gallup poll, for example, like many polls before it, ranked car salesman as the least trustworthy profession.[7]

Adding to the tension was the system used by dealers to sell cars. My father would sometimes tell me about "taking an up," which means meeting a new customer who walked into the showroom for the first time. In a common practice, salesmen added their names to a list and took turns when each new customer, or "up," arrived at the showroom.

While it is almost impossible for me to know what exact techniques my father used to keep food on the table, well-documented and questionable business tactics used by car salesmen at that time included "the blitz," where teams of salesmen pressured customers in small rooms, so-called plain packs, or inflated charges for dealer preparations of the car, high rates on financing that often included kickbacks, and the notorious "bait and switch," in which a customer is lured into the showroom with a promise of a good deal only to be forced to accept a worse deal on another model. [8]

Women were often targets of their chicanery. "Since Henry Ford's Model T first rolled off the assembly line, automobiles have been called upon to perpetuate the dominant ideological positions and reinforce power differentials," writes Chris Lezotte in "What Would Miss Daisy Drive? The Road Trip Film, the Automobile, and the Woman behind the Wheel." [9]

Nowhere were these power differentials more clearly displayed than in the showroom. Women not only bought their own cars early on, but they were also a little-acknowledged force in the purchasing decisions for family cars, a buying power that was part and parcel of their newfound freedom to participate in the wider economic and political world. Well-known suffragist Elizabeth Cady Stanton, for example, reportedly encouraged women to assert their economic power with the admonition "Go out and buy!"

Friedman notes that data from Ford in the 1920s shows that women most likely had "a better than 51 percent of the say when the purchase of a car is decided." [10] General Motors would adopt the slogan "A Car for Her" in the 1920s, promoting the now ubiquitous two-car lifestyle. The ad campaign had a corollary aimed at farm women: "Every farm should have two cars," read a 1923 ad for the Chevrolet Four Passenger Coupé. The two-car strategy, subtly aimed at women, was credited with helping General Motors pass Ford as the nation's largest automobile manufacturer.

"By the 1920s, advertisers targeted women as the shoppers at the center of the world of consumption," Cynthia Wright writes in her essay on women and consumption, "Feminine Trifles of Vast Importance." [11] So much so that by 1929 the *Journal of the American Advertising* thundered: "The proper study of mankind is man, but the proper study of markets is woman." [12]

Yet despite the early presence of women as car buyers, it was a long road to acceptance for women customers in the showroom. The lack of technical know-how and early legal impediments that kept women from acquiring credit on their own put the power differential squarely on the side of the salesman.

As early as 1926, Chevrolet instructed their salesmen in the proper way to pressure women into purchasing an automobile complete with this oily script. "Naturally, every woman will want to talk over so important an item as the purchase of the family automobile," the script read. "Your husband will probably be even more interested than you are in many of Chevrolet's desirable features especially its mechanical superiorities . . . He will heartily approve your choice when you tell him that Chevrolet uses a banjo type rear axle. You sign the order now and when I deliver your car tomorrow, I can explain these quality features to him."[13]

Consider the sales strategy used in the 1950s by Hodges Auto Sales, a Michigan Dodge dealer with a catchy slogan: "See Hodges for Dodges." Every day salesmen sent out postcards to residents that alerted them that a salesman would soon phone them. The salesman had one agenda during that call, according to Robert Genat, author of *The American Car Dealership*, to "qualify" prospects with leading questions such as: "Do you folks want your new car this week or next?" After this provocative question, the salesperson sought to discover the wife's favorite color by visiting her at home. "With the 'right' color determined, the salesman would select a car that most suited the prospect and in the color the wife favored." According to Genat, a salesman brought the car to the home, where the wife could approve it, but the deal was done with the husband back at the office.[14] Genat says that salesmen were instructed to demonstrate to women vehicles with lots of options—known as loaded vehicles—cars with options such as power steering, power brakes, and subtle comforts such as tissue dispensers and vanity mirrors. To further appeal to women customers, dealers often handed out copies of *Handbook for the Woman Driver*, published in 1955 and written by the automotive editor for the ladies' magazine *Good Housekeeping*, Charlotte Montgomery. Montgomery begins her book by saying, "A woman has a very special feeling about the car she drives," before adding routine tips on tire rotation, advice on keeping the kids entertained on long rides, and a full-page illustration on how to parallel park.

Later, as showrooms became more ornate, sometimes including grand pianos and well-polished floors and vehicles, sales manuals reminded salesmen that the reflective surfaces in the showroom could easily reveal an eye roll or other ways of shrugging off a woman's comments. According to one historian, it is all part and parcel of the general disregard that the industry holds for women.

"The story of American Automotive history reveals a long-standing pattern of according men respect and women disdain," Katherine Parkin writes in *Women at the Wheel: A Century of Buying, Driving, and Fixing Cars.* "Women across the century bought cars in what observers described as a male space, with one marketer describing it as having a 'locker room' mentality." As Parkin concludes: "The people and the place were both structured around male consumers. Salesmen believed women who walked through the doors to be naive, misguided, or looking for love."[15]

Sales plans for car salesmen were replete with prizes called "spiffs," point systems, and games. In 1963 one Chevrolet official said, "We have less than 7,000 dealers and nearly 10,000 compensation plans."[16] The average salesman may have made upward of $12,000 in the 1950s, but I doubt my father ever got close to that. He was a man tasked with selling the American Dream even though he was never able to obtain it. We lived paycheck to paycheck, sometimes without heat and at times with very little food, but remarkably there was always a good deal of beer and plenty of money for cigarettes. His shoes had holes filled with cardboard, and my underwear was often tattered. It was subsistence living, albeit with very nice new cars.

At times I can remember very little about my dad, but I sure do remember what he drove.

TWO

My Mother's Chevy Convertible and Sexual Freedom

Someone should write an erudite essay on the moral, physical, and esthetic effect of the Model T Ford on the American nation. Two generations of Americans knew more about the Ford coil than the clitoris, about the planetary system of gears than the solar system of stars.

—John Steinbeck, *Cannery Row*

If the invention of the car was most accurately ascribed to Daimler and Benz of Germany, the mass production and mass marketing of the automobile was invented in my own fertile Midwest. Inspired by the carcass disassembly process of the Chicago Stockyards, Henry Ford devised the vast assembly lines that would come to define Detroit. His Model T, nicknamed "Henry's Lady" or "Tin Lizzie," was introduced in 1908, and his assembly lines lurched into action in 1913 in Highland Park, Michigan.

Far from the complicated, sophisticated industry we now think of when we contemplate the auto industry, Detroit then was populated by a group of smart, determined tinkerers; quarrelsome young men capable of working with their hands, men determined enough to work in unheated garages during long midwestern winters, men strong enough to lift the engines and turn the cranks that started early models.

"The auto industry was born in a masculine manger," Virginia Scharff writes in *Taking the Wheel: Women and the Coming of the Motor Age*. And, perhaps because of this, when most people think of the automobile, they think about men.

To the extent that people link women and cars, a kind of pinup mentality rules: think bikini butt shots on hot rods, half-naked women on the hoods of cars gracing auto shop calendars. (Just google women and cars if you doubt me.)

But beyond the men-and-machine myth lies a different story. In the short, just over hundred-year history of the automobile, one thing has been constant: women have always loved their cars. Wealthy and poor women alike took to the open road with stylish abandon.

Cars were a big part of my childhood. They shaped who I am today in the same way that they shaped and changed the collective experience of women in multiple ways. The car changed the way women worked and lived, restructuring the status of women and dissecting their hometowns with a vast networked highway system, one that delivered them to the suburbs, where their work of shopping and chauffeuring would be concealed behind the high gloss of a station wagon.

The automobile created the suburbs with all the good and the bad that that brought to women's lives. There, in surprising ways, the car enslaved women even as it liberated them. Where once "women's work" was visible to all—say, in the vegetable garden or as laundry hanging on a line—much of women's work in the suburbs took place within the confines of the car as the almost constant shopping and chauffeuring of children to activities outside the neighborhood became both necessary and routine. Cars created the possibility for connection and commerce at the same time that they created a unique form of isolation. Women reaped enormous rewards and paid an enormous price.

The car, at least in the first instance, seems to have liberated my mother.

She was striking—five foot nine and thin—and if she wasn't exactly beautiful, she certainly gave the impression of great beauty. Perhaps it was her clothes, which were meticulous and impractical: bouclé suits in winter and, in summer, starched skirts with dozens of tiny pleats she hand-ironed, only to watch their sharp edges wilt in the unrelenting heat and humidity.

Her car was as fashionable as her clothes and equally impractical. She drove a used light blue Chevy Impala convertible from the 1960s. The car had white vinyl bucket seats, and the shift was on the steering wheel column. She drove fast, and if she had to stop on short notice, she flung her arm across my chest in the impromptu seatbelt move common to mothers of the day.

We made our getaway from my father in that old Chevy one cold winter night. Legal scholars have often noted that driver's licenses and access to cars played a key role in the lives of abused women, who were suddenly free to seek help. [1]

Violence was one of the few grounds for divorce in 1964, and my mother's divorce decree, which I discovered years later, does refer to several instances of "violence and cruelty." My best guess is she just hit her breaking point after years of dysfunction and abuse. She took me with her. I was five years old, and it was 1965. We left behind my older brother and sister. They were in their late teens, and I suppose they got to choose.

I remember sitting in the driveway, the engine running to keep us warm, my breath blowing a light fog onto the window. My father came out to the car. I rolled down the window, working that old-style crank hard, and he handed me a sleeve of saltines.

"Here," he said.

The Chevy lurched into gear, spitting gravel as we turned out of the driveway and headed across town to the one-bedroom bungalow my mother had rented. The thin convertible roof leaked cold air, the car bucked as the engine slipped into gear, and the crumbs gathered in my lap.

I drank my first sweet cup of coffee that night out of an old lilac melamine cup. Munching on crackers and sitting on the floor of the tiny one-bedroom house, I thought it was going to be an extraordinary adventure. Although I didn't know the word at the time, it was exhilarating.

As a child, I didn't appreciate what a desperate act it was. My mother was the only divorced woman in town. She made little money as a secretary in the athletic department at the high school, and my father's support came in fits and starts, probably because his own success was so tenuous. Sometimes he would hit it big with a new model or sales promotion, but other times a plant closure or a strike at a factory would slow sales for months. During the long winters, it wasn't uncommon to come home and find the heat or electricity shut off because she'd been so late paying a bill.

Adult lives are always a source of mystery to children, and my mother's life was no exception. Our small industrial town had almost no social outlets for a newly divorced woman. My mother was Catholic, and she was either excommunicated from the church or simply felt too uncomfortable to continue to attend.

I remember one picture of her sitting at a restaurant with some of the other secretaries from the high school. She was wearing a white dress and a peach silk coat, her cigarette waving in the air.

She used to listen to high school basketball games on the radio on Friday nights. She knew many of the boys personally—those who were acting out or in danger of becoming ineligible to play because of poor grades were sent to her office during homeroom—and I would hear her calling out encouragement to them as she listened. She clapped her hands at each free throw they hit. She knew well enough that an opportunity was not something anyone could afford to waste.

On cold winter nights, her old Chevy would idle in the drive as she put on her makeup and waited for me to fall asleep. I didn't know where she went at night, only that she was out.

Sometimes I would awaken to voices in the living room. Whenever she had a man visit, I was instructed to call him "Mr. X." One night I awoke to a tall fellow whirling my neon Hula-Hoop on his hips in the dark, and for the longest time that was my vision of sex—that lurid green Hula-Hoop twirling in the nighttime and the Chevy idling in the driveway.

✳

We take this fact for granted now, but the car permanently altered the terrain of romance. It liberated rural and small-town women from the farm and introduced them to the pleasures of the world: boyfriends, parties, libraries, and soda fountains. It took ladies in smart outfits to the country club and matrons in pearls and black velvet to the opera. Where once courtship had been confined to the parlor or the front porch, romance now took place far from parental supervision in a car on a dark country road. Next to the birth control pill, perhaps no other technology has so greatly influenced the sex lives of women as the car.

Ironically, the push for the automobile was brought about in large measure because it was difficult for women to preserve their modesty, their feminine virtue, and their bourgeois status on crowded public street cars. Back then, the car was touted as a solution to the specialized needs of women, but soon the car itself would become a threat to women's modesty, chastity, and safety.

Almost as soon as they hit the road, cars were used for sex. Early models with long, wide running boards made for novel locations for trysts when covered with pillows and blankets. Later models boasted reclining front seats that typically slept two people in comfort. Henry Ford, a man not ashamed to foist his conservative views on customers and employees alike, was so aghast at the notion of rampant sex taking place in his Model T that he tried to limit seat length to thirty-eight inches so as to make it downright uncomfortable, if not impossible, to make love in Tin Lizzies.[2]

No matter. Sex in cars persisted. Often aided by better design.

As the automobile became more common, it also became more comfortable. In 1920, Americans owned more than twice as many horses as cars, but by 1930 the reverse was true, approaching on average one car per household. By then most cars had become enclosed, and passengers sat lower and were partially hidden by side panels.

Progressively, cars were made more welcoming for sex, although it is unclear whether this was intentional or not. The 1925 Jewett, for example, had a fold-down front seat that comfortably slept two people so long as they were less than six feet tall. The president of the Nash Automobile Company at the time was said to have quipped that several of his models—including the Statesman—were "young men's cars" since their seats so quickly and completely folded down for sex.

By the mid-1930s, cars had become so prevalent and important to American culture that President Herbert Hoover's 1933 Commission on Social Trends warned that the automobile had the power to change habits of both thought and language. And it did so with a vengeance.

While the language to describe car parts often referenced the female body—headlights were buxom and grills were either bust-like or vaginal—the cars themselves were often described as piercing and powerful with horsepower as a proxy for male potency. Ford, for example, in ads for one of its bestselling vehicles touts its "power stroke" engine. A man drove a car hard, smoothly shifting gears for optimal performance and the ultimate experience. If gender is, as some scholars argue, largely performative, the car was a perfect set piece.

The theme of sexual dominance as expressed through the automobile would play out over decades in ads, greeting cards, and literature. Jay Gatsby's cream-colored Rolls-Royce in the popular 1925 novel *The Great*

Gatsby, for example, is described as "swollen" and "monstrous" in length with a "labyrinth of wind-shields that mirrored a dozen suns."[3]

Similarly, consider one of the main characters in the novel, Jordan Baker. Baker was named after two swanky car companies of the time, the Jordan Motor Car Company and the Baker Motor Vehicle Company, which some literary critics read as a sign that Baker was a "fast" woman open to sexual adventure.

Wealthy ladies, especially socialites like Baker, were among the first women to own cars. And yet the early cars demanded strength both to start and to steer, and footmen or other servants soon found themselves in the role of chauffeur—a term originally derived from the French *chauffer*, meaning "to stoke or fire up." The word probably refers to the job of early drivers, who cranked the motors of the first cars.

But of course, the engine of the car wasn't all the chauffeur might fire up, or at least that's how the pulp fiction writers of the time imagined it. In the colloquial, *chauffer une femme* meant "to make hot love to a woman." The chauffeur threatened class and gender lines by leaving women alone with men of another class in an enclosed space and outside the home for the first time.

As a result, the car became both the source of and the setting for sensational, melodramatic tales in which the woman lusts for the explosive power of the automobile. In a 1905 Hearst *Motor* magazine serial, "The Wonderful Monster," a fictional character named Lady Beeston waxes poetic about her car:

> To think of "The Monster," as she called it, was to long for it. That great living, wonderful thing with its passion for motion seemed to call and claim her as a kindred spirit. She wanted to feel the throb of its quickening pulses; to lay her hand on lever and handle and thrill with the sense of mastery; to claim its power as her own—and feel its sullen-yielded obedience answer her will.[4]

To use another literary example, consider the work of Edith Wharton, herself an early adopter of the automobile and a great car enthusiast who thought the car created "an immense enlargement of life." Undine Spragg, the protagonist in Wharton's 1913 novel *Custom of the Country*, is so taken by a drive in the country that she feels a "rush of physical joy that drowns scruples and silences memory" in the car.[5]

Spragg was not alone. For much of the twentieth century, for people across a broad swath of economic classes, cars were "generally considered the best and most private option for sexual privacy," according to historian Beth L. Bailey.[6] In the 1920s, a popular valentine read YOU AUTO BE MINE. Other popular variations included sexually loaded references to spark plugs, starter cranks, and drive shafts. In the 1930s, a student publication at the University of Chicago "recommended the automobile 'whenever possible' for sex."[7]

With the growing popularity of the automobile, romance quickly moved to the back seat. "In twentieth-century America, courtship became more and more a private act conducted in the public world. This intimate business, as it evolved into 'dating,' increasingly took place in public places removed, by distance and anonymity, from the sheltering and controlling contexts of home and local community," Bailey writes. "Keeping company in the family parlor was replaced by dining and dancing, Coke dates, movies, and 'parking.'"[8]

Dating had gradually, almost imperceptibly, become a universal custom in the United States. Whether cars were the cause of this vast change in mating styles or simply fuel for an already changing romantic landscape, which included breakdowns in class structure and increased industrialization and urbanization, is not clear. By the early twentieth century, however, the car had given young people both mobility and privacy and therefore had become a driving force for romantic and sexual liaisons. "Parking" became synonymous with a romantic evening in a car. "Lover's lanes abounded in parks and off lesser streets and roadways in and around most communities," David L. Lewis writes in "From Rumble Seats to Rockin' Van."[9]

Unlike earlier decades, when a man might "call" on a woman at home in the presence of her parents or a chaperone, dating now meant going out and spending money. In this new system, capitalistic rules and economic language applied. Men asked women out on dates and paid for the pleasure of their company with the sometimes subtle and sometimes not-so-subtle subtext that the evening's activities were provided in exchange for sexual favors. Cars added fuel to this new dynamic, putting men in the driver's seat and women in a submissive role on the passenger side. "What men were buying in the dating system was not just female companionship, not just entertainment—but power," Bailey concludes. "Money purchased obligation; money purchased inequality; money purchased control."[10] So rampant was the accumulation of sexual conquests, young cads were known to call their cars the *Mayflower*, referencing how many women came in them.

Songs from the early part of the twentieth century warned of the various pleasures and dangers of getting into a car with a man. They included "Take Me Out for a Joy Ride" and "Keep Away from the Fellow Who Owns an Automobile (The Cad!)." Later songs laid clear the emotional and sexual deal required of a woman who took a ride in a car. In Bruce Springsteen's 1986 hit, "Thunder Road," he offers redemption under the hood with the caveat that "the door is open, but the ride ain't free."

As in so many gendered equations, men had so much to gain and women so much to lose in a car ride. Women seeking or submitting to sex found that they risked both their reputations and unwanted pregnancies. Men, on the other hand, stood mainly to gain in peer standing from sexual encounters. "Getting into a car added a dimension to men's relative power in sexual encounters," Virginia Scharff writes, "since more men than women drove, more men than women owned cars, and even where both were qualified, men usually took the wheel."[11]

And what a sad story it was when a woman got into a car with the wrong man. In one incident from 1947, a male sophomore at the University of Michigan was driving a nineteen-year-old woman home from a party when he turned onto a deserted road and beat and raped her. Arrested by the police, he was then suspended from the university. But get this: she was too. The university deemed her conduct neither a credit to herself nor to the university. Her crime: getting into the car. One 1967 article in *Cosmopolitan* titled "Do Women Provoke Sex Attacks?" argued that they do, even if it is not always conscious. It was a prevalent attitude throughout most of the twentieth century.

In part because of concerns about sexual violence and to tamp down on sexual behavior in general, colleges and local governments sought to limit automobiles on college campuses and passed laws that made stopping and parking constitute reckless driving. Young girls dragged out of vehicles by police patrols sometimes ended up in juvenile court, where one judge called cars "houses of prostitution on wheels."

If these proposals were effective, you couldn't prove it by the activities in my high school in the late 1970s. By the time I was a teenager, car culture was in full force. And it was overwhelmingly male. Boys rebuilt cars in their front yards and were required to take shop in school—a class not offered to girls.

On Saturday nights we "scooped the loop" in our newly waxed and buffed cars. The more adventuresome gearheads would meet at a local gas station to remove the mufflers from their cars, add straight pipes, and proceed to burn rubber down Genesee Street, the town's main drag. By the end of the night couples paired off for teenage petting down by the lake. More than a few couples would end up married, like Bob Schultz, who spent his entire thirty-day leave from the Air Force driving in small circles downtown until he met his wife, Marilyn. He later told the local paper, "Subconsciously, the only way I wanted to find my wife was cruising."[12]

Other alternatives for amorous adventures in cars opened up as well. The drive-in by the highway was a big destination. "The drive-in theater almost seemed to have been created for sex in cars," David L. Lewis writes in *The Automobile and American Culture*. The first drive-in theater opened in June 1933 in Camden, New Jersey, and it could hold four hundred cars aligned in seven rows. Successive drive-ins grew to multiscreen venues, with some offering as many as nine screens with eighteen viewings per evening. Many establishments were equipped with individual heaters and air conditioners. Drive-ins quickly gained a reputation as "passion pits," with some patrons suggesting that the show in the cars was better than the show on-screen.

Other, more adventurous, teens experimented with even more dangerous and exciting possibilities than visiting the drive-in, such as sex in cars moving at high speeds. Road head—fellatio performed on someone who is driving—has been extolled as the pinnacle of vehicular sex acts in *Esquire*, caused a humorous crash in the film *Parenthood*, was romanticized in the hip-hop song "Homie Gets Some Road Head," and tragically caused the death of a bicyclist in Florida, in 2015, who was killed by the driver of a pickup who was enjoying oral sex from his passenger.

The automobile, as President Hoover feared, did change habits of both thought and language, and nowhere is that more obvious than in aggressive bumper stickers and vanity license plates that stress the long-standing link between sex and cars. One particularly egregious bumper sticker reads FOUR DOORS FOR MORE WHORES. A familiar sticker from my youth stated ASS, GRASS, OR GAS: NOBODY RIDES FREE.

In the past several decades, license plates have been used to elicit sexual partners and to signal sexual preferences and identity. Vanity plates boast any number of combinations but amongst the most popular are SEXY, NO WIFE, and GAY. Showing rare good taste, the California Department of Motor Vehicles has reportedly denied such requests as ONW2BYB, an acronym for "On My Way to Bang Your Bitch," and DICKOUT, which needs no translation. The state of California has been known to have a waiting list for license plates with the letters GAY. Back in 1978, however, more than one hundred Iowans randomly assigned a GAY plate paid a $4 exchange fee to get rid of them.

The future of sex in cars, however storied its past, remains unclear. Teenagers now take advantage of homes where both parents work, and college students can more easily visit each other's dorm rooms. Increasing demands on the interior of the car for safety accommodations, such as air bags, and the need to make the vehicle as aerodynamic as possible for fuel efficiency have made sex in the constricted interior of the twenty-first-century car limiting, to say the least. "Even the most determined, ingenious and acrobatic of lovers will find them an all but impossible challenge," one auto critic wrote of recent models. [13]

Some have predicted the end of sex in cars altogether, calling it anachronistic and unnecessary. Others predict that the driverless car that requires no human intervention will open a whole new chapter of sexual exploits in cars. One can imagine, after all, a very different commute if your entire ride is automated. More than one prototype of an autonomous car shows side-by-side reclining seats that resemble luxury twin beds pushed together, a design reminiscent of early model cars with fold-down front seats.

As for my mother, I have no idea what she did in her convertible in the 1960s, but I know this for certain. That car radiated romance—just as sure as it smelled of Chanel No5—and presumably, since we shared a bedroom in our small house, her car was used for sex more than once.

There used to be an old joke in the Midwest that it didn't matter what kind of car you drove, your last ride was always in a Cadillac. That was certainly true for my mother. By the time I was growing up in the Midwest, Cadillac had cornered the market for hearses.

I had never ridden in a limousine before the frozen February morning one pulled into our driveway to take my family to my mother's funeral. I was ten, and my mother had died in my arms just days before; a kind of shock had come over me. I was tingling on my right side, a symptom

I didn't disclose to anyone and one that would linger into adulthood any time I felt overwhelmed by events.

I don't know whether my mother succumbed to stress or a complex cocktail of prescription drugs and alcohol. She took a heady mix of drugs, including Darvon (now banned), a drug known to create an irregular heartbeat, and birth control pills, which in their earliest forms were linked to blood clots. And she smoked.

I was alone with her when she died. I knew she wasn't well. When she walked up hills her breathing was labored, and her drinking had left her so weakened that she sometimes did not get out of bed on weekends. So, when I saw that she was unable to get out of bed that morning, I refused to go to school, as I sometimes did when she'd had too much to drink. She complained about it but fell back asleep. I had made myself a frozen waffle and turned on the television when I heard her head hit the wall. I was fast calling 911, but they were unable to revive her. I sat near the corner of the room as the bed shook from the defibrillators, my small dog trapped beneath the bed snarling as each vain attempt was made.

My mother had always loved yellow roses. The yellow rose was the flower and symbol of her saint namesake, Saint Rita of Cascia—appropriately enough, the patron saint of desperation and hopeless marriages.

Saint Rita was from a warring family in Italy and was likely an abused wife. It was thought that her prayers changed the violent ways of her husband and brought an end to a decades-long feud with another village family. After her husband's death, Rita petitioned three times to join the convent and was eventually admitted. She is credited with several miracles during her time there, one of which was making a rosebush bloom in winter in the frigid hills of Umbria. Thus, petitioners often leave roses and rose petals at the memorial to her at the Basilica of Santa Rita de Cascia.

My sister, in some mixture of anguish and guilt at the death of our mother, decided that her casket would be covered in yellow roses. It was the kind of dramatic gesture we were good at: the carpets were threadbare, the electric bill was overdue, and my mother's death meant the loss of one half of our family income, but no matter. We would have a six-foot blanket made of hundreds of roses to cover her casket in the dead of a midwestern winter. It was a small miracle in and of itself.

My final encounter with my mother's suitors was at her funeral. The pallbearers—six tall men from the school where she worked, some of whom, at one time or another, might have answered to the name of Mr. X—lifted her casket high onto their broad shoulders and delivered her lifeless body to the back of a flower-festooned Cadillac. Later they would slide the casket into the frozen ground. That's what I remember. Hundreds of small yellow tea roses wired into one great blanket, their petals quivering in the frigid winds as the casket was lowered carefully by some nice-looking men. I do believe my mother would have been pleased by the spirit of the whole thing: beautiful flowers, good-looking men, and a fine, clean car for her last ride.

THREE

My Sister's Mustang: A Fashion Accessory Made of Steel

Boomers responded to the Mustang because it was a beauty, it had power, and it had an identity. The car means something to them.

—Lee Iacocca

I n the late 1960s the twin themes of free love and civil disruption were weaving their way through American culture. War was raging in Vietnam, and it was not uncommon to see the US flag at the high school fly at half-mast when one of our boys serving overseas was killed.[1]

Across the country, race riots erupted in several major cities, including the great midwestern behemoths of Chicago and Detroit. Riots erupted in our small town, too, down near the housing projects where most of the Blacks lived. A few Molotov cocktails were tossed into the street, and the Cat and the Fiddle, a local bar for Blacks, was forced to close its doors.

Just south of my hometown, the Chicago Seven—a group of well-known antiwar activists accused of planning to incite a riot at the 1968 Democratic

National Convention—were famously bound and gagged at their own trial. And, as if to prove that everything had reached a fever pitch, a record 54,985 automobile deaths would be reported in 1969.

In the summer of 1969 my sister, Sue, was nineteen, and I was ten. She had long black hair that she often wore teased into a sky-high beehive. She had perfect handwriting, which she scratched out with a fountain pen on thin-lined notebook paper. During that time, she was a commuter student at the University of Illinois Chicago, and each morning she would hop in her little Mustang and head down the toll road to school.

My sister's Mustang—a red convertible with a white interior that our father had bought for her used—was a fast car, but she drove it slowly. She was absentminded and unhurried in the way of many great beauties. She would play with her hair while she drove, shifting gears and flicking her hair in a kind of rhythmic beat. Flick, shift, flick, shift.

After my parents' divorce, my sister continued to live with my father, but when I was in grade school, she would often zip across town to pick me up at my mother's house for a small afternoon adventure.

In the Mustang, my sister and I escaped from mundane streets, heading out to fancier suburbs to explore things our small town lacked, including bookstores, shopping malls, and the vast oasis that was the giant department store Marshall Field's on State Street. In that way we were like early women motorists who used their automobiles to explore newfound pleasures such as roadside picnics and soda fountains. Indeed, one poor rural woman told an investigator in 1920 that she went without running water in order to pursue more cosmopolitan pleasures, explaining what she thought was obvious about her priorities: "Well, you can't go to town in a bathtub, can you?"[2]

My sister and I did the same. We went joyriding, searching for freedom and pleasure that was missing in our home, which was rich with automobile parts and lacking in almost everything else. My sister often took me to the beach—that sliver of sand down near the factories in my hometown—

where we sunbathed under the shadows of the smokestacks and dodged the dead fish that washed up on the polluted shores of Lake Michigan.

Sometimes when we wanted to go out to lunch, we would eat a grilled cheese sandwich at the counter at Woolworth's or head north to a drive-in restaurant called A&W, where a carhop would attach a tray to the window of my sister's car and bring us ice cold mugs of root beer and a sandwich called a Belly Whopper.

Some days we would go clothes shopping, fingering materials and trying on clothes we couldn't afford under the watchful eye of snooty salesclerks. My sister loved clothes, and her tall, thin body made her a perfect mannequin for the minidresses and bell bottom styles of the 1960s. Her car—a bright and happy little dream built for idealistic and irreverent Baby Boomers on a budget—served as a kind of fashion accessory, a perfect backdrop to her mod prints, lace minidresses, halter tops, and fringe suede jackets. A well-known ad at the time encouraged women to "Wear a Mustang to Match your Lipstick."[3] And while my sister never took that fashion advice, she certainly always looked good in her little red car.

From the earliest days of the automobile, women were expected to look presentable as they traveled on mud-filled rutted roads in vehicles exposed to the elements. As a result, clothes and cars have been stylish mates since the beginning of the life of the automobile. Early terms for car parts were derived from women's clothing. The *bonnet* covered the engine, and *skirts* lined the sides of the automobile. *Dagmars* were decorative chrome protrusions on the front bumpers of mid-1960s Buicks and Cadillacs, named after the conical bras worn by a voluptuous Hollywood star named Dagmar, aka Bubbles. Hood insignias, such as the Chrysler crest and the Mercedes grill star emblem, were referred to as *jewelry*.

Inside the car, the clothing women wore was designed to match the rugged and sometimes dangerous conditions that met the first drivers and passengers. In 1904, for example, just a handful of years after the invention of the automobile, Saks Fifth Avenue published a one-hundred-page catalog of coats, hats, goggles, gloves, hoods, and lap blankets for the stylish driver and their passengers, giving it the clear and somewhat earnest title "Automobile Garments and Requisites: Imported and Domestic Models for Men and Women." Page after page describes garments ranging from pragmatic to luxurious—those that went well beyond the requirements needed to travel safely; others that skirted perilously into the forefront of fashion, including a full-length mohair cape whose long pile was described as "dense and fur-like" and whose "sleeves, fly front and shoulders are appliqued with white kid leather out of which diamond shaped panels are cut, so that the plush may show."[4]

Looking good was part and parcel of what was expected of the early female motorist. "The woman who runs and cares for her own car should dress simply and plainly in a style that will enable her to appear neat and smart under all circumstances," Mary Walker Harper wrote in 1915 in the *Ladies' Home Journal*.[5]

Harper suggested that the smart motorist wear "a tailored gown of black satin with a white collar, a small, close-fitting white hat with a black veil of the finest mesh to keep the hair neat, white washable gloves, with several extra pairs in the car handy for use, and common-sense shoes with low, broad heels." Harper goes on to explain. "By these means you will be comfortable, a credit to your sister motorists, and a joy to passers-by to behold—calm, self-contained and sure of yourself."[6]

From the very beginning, motoring was a performative act, and women were expected to play the part. Dorothy Levitt, a groundbreaking early female motorist and race driver, for example, recommended keeping a little drawer of accessories in the car, an early forerunner to today's glove compartment.

"This little drawer is the secret of the dainty motorist," she wrote in her 1909 book *The Woman and The Car, A Chatty Little Handbook for Women Who Motor or Want to Motor.* "What you put in it depends upon your tastes, but the following articles are what I advise you have in its recesses. A pair of clean gloves, an extra handkerchief, clean veil, powder-puff (unless you despise them), hair pins and ordinary pins, a hand mirror—and some chocolates are very soothing sometimes." She also thought that for the woman who traveled alone a small revolver—she suggests a Colt—might be "advisable," and that a dog can make for "great company."[7]

Wealthy women, like Leavitt, were amongst the first to gravitate to the car, and it was wealthy women who first exhibited the conspicuous consumption, including luxurious clothing, that travel by car signaled in the beginning of the twentieth century. For example, Pulitzer Prize–winning author Edith Wharton, in her well-received 1908 book, *A Motor Flight through France*, catalogued her travel in her chauffeur-driven twenty-four-foot 1904 Panhard-Levassor, a car that her friend and fellow author Henry James called her "vehicle of passion."

"The motor car has restored the romance of travel," proclaimed Wharton, who argued that the car had freed travelers from the strict schedules of the railroad and recovered such pleasures as "taking a town unawares, stealing on it by back ways and un-chronicled paths, and surprising in it some intimate aspect of past time."[8]

Several years after the publication of *A Motor Flight through France*, World War I would begin to make driving routine, even necessary, for women. But the road was neither straight nor smooth for women motorists. As Virginia Scharff has written in her history of women and cars, *Taking the Wheel: Women and the Coming of the Motor Age*, "Simply getting into a motor vehicle made women too prominent to escape notice; actually, taking the wheel made them too astonishing to escape controversy."[9]

Women were at once deemed too nervous, too timid, too weak—altogether constitutionally unfit to drive. Driving also shook long-held social structures that required women to tend to home and hearth. Indeed, a 1916 article from *Motor Age* recounted one of the many perceived problems with women driving—lowered egg production—as gallivanting women were not home minding their "biddies," as baby chickens were called. The ladies were not having it. "Women of the central Illinois farms emphatically denounce the libel that the egg industry has gone to pot because they are spending most of their time scudding across the country in motor cars and neglecting the poultry yards," the author reported. [10]

Historian Michael Berger argues that the tension was palpable.

> "The implied conflict between women's desire to expand their social and economic horizons and society's fear that such a development would lead to an abandonment of women's traditional roles in society was very real. For although often presented in a humorous context, folklore concerning women drivers, and the accompanying negative stereotype emerged for very serious social reasons. They were attempts to both keep women in their place and to protect them against corrupting influences in society. Once the automobile came within the financial means and mechanical abilities of most Americans, its potential for altering social and economic values was quickly realized." [11]

More economical cars and the changing mores of the Roaring Twenties combined to liberate women from airtight social strictures, changing both the way they dressed and what they drove. Flappers, for example, were often depicted dangling over long cars with even longer strands of pearls, wearing small cloche hats and daringly baring long arms underneath short sleeves.

Perhaps the most famous vehicle of that era was Jay Gatsby's Silver Ghost. [12] The Ghost, driven by Daisy Buchanan, would famously kill her husband's working-class lover, tearing off her breast and leaving her dead body in the street—a scene that highlights not only how dangerous early vehicles were to pedestrians but also the class distinctions that early automobiles embodied.

And then there's Jordan Baker, a character named after two car companies, known for marketing glamour and comfort in their autos to women: the Jordan Motor Car Company and the Baker Motor Vehicle Company. The Baker Electric was a popular electric model produced in the 1920s, a car whose "Eden-like ads"[13] featured women in petticoats and little girls in white dresses frolicking in a garden with butterflies. An ad that not so subtly proposed that the car was a perfect tool for mother-daughter bonding among the wildflowers.

Beyond its portrayal in *The Great Gatsby*, the Jordan Motor Company is best known for having the first ad that linked sex, women, and nature in a robust, poetic attempt to sell cars without mentioning any vehicle attributes. In 1923, the company's "Somewhere West of Laramie" ad appeared in the *Saturday Evening Post*:

> Somewhere west of Laramie there's a bronco-busting, steer roping girl who knows what I'm talking about.
>
> She can tell what a sassy pony, that's a cross between greased lightning and the place where it hits, can do with eleven hundred pounds of steel and action when he's going high, wide and handsome.
>
> The truth is—the Jordan Playboy was built for her.
>
> Built for the lass whose face is brown with the sun when the day is done of revel and romp and race.
>
> She loves the cross of the wild and the tame.
>
> There's a savor of links about that car—of laughter and lilt and light—a hint of old loves—and saddle and quirt.
>
> It's a brawny thing—yet a graceful thing for the sweep o' the Avenue.

Step into the Playboy when the hour grows dull with things gone dead and stale.

Then start for the land of real living with the spirit of the lass who rides, lean and rangy, into the red horizon of a Wyoming twilight.

The ad marked the beginning of the long and storied association between cars, a rugged outdoor spirit, and the now trite, but still successful, selling point of sexual satisfaction heightened by horsepower.

Women responded to the "Somewhere West of Laramie" ad with an unexpected, robust, and randy enthusiasm. "I do not want a position with your company," one woman wrote in a letter to the Jordan Company. "I just want to meet the man who wrote that advertisement. I am 23 years of age, a brunette, weigh 120 pounds and my wings are spread. All you've got to do is say the word and I'll fly to you."[14]

Despite the enthusiasm voiced by this young woman, things were changing. By 1929, according to automotive historian James Flink, "Everyone who could afford to buy an automobile already owned one, and the average life of a passenger car was then estimated by the industry to be seven years." The result was that sales and advertising were necessary to create an aura around a car that had little to do with meeting basic transportation needs.[15]

Car companies, for example, began to suggest that a car could both set a mood for an occasion and complement a woman's personal style. Car companies overwhelmingly turned to fashion and style to stoke sales with the woman as their target audience for elegant ads. According to Roland Marchand in *Advertising the American Dream*, the REO Motor Car Company bragged that its car finishes "subtly recaptured the season's new fall tones from Paris designers."[16]

The ads proclaimed, "A woman's car today should be her most charming background."[17] Another auto company, Paige-Jewett, advertised automobiles in fourteen body types, promising a car that would "match milady's mode."[18]

"Beauty is what sells the American car," one auto executive told *Time* magazine. "And the person we're designing it for is the American woman. It is the woman who likes colors."[19]

Throughout the 1930s and 1940s, car shapes and fashion silhouettes moved in synchrony with one influencing the other. In 1939, for example, to introduce its new "sealed beam" headlights, Ford asked designers such as Jay Thorpe to create "headlight hats" for models to wear at car shows, including one hat that consisted of "a tasty brown pancake of nutria fur, and another the size of a Christmas poinsettia."[20]

According to a 1940 article in *Good Housekeeping*, "Hats were cumbersome. So were auto tops. Skirts were clumsy. So were chassis. Corsets were uncomfortable. So were back-seat springs. Hampered by their trappings, ladies moved sedately. The bulky gasoline buggy followed suit. It stalled. It puffed some. It proceeded uphill—if at all—laboriously and with noisy complaint. Today all that has changed. Automobile tops are curved to flow with the wind. So are hats. Hoods are slim and elongated. So is the modern torso. 1941's cars and fashions are well designed for comfortable, efficiency—tapered to sleek lines and geared to proceed in high fashion."[21]

The charismatic connection between the female form and car stylings was not always benign. Buicks were thought to have "bosom" bumpers, and observers were quick to pick up on Buick's ring-shaped hood ornament pierced by a "flying phallus." Some critics even believe that the flop of the Edsel had much to do with its vaginal-like center grill with its teeth-like design that made it seem more intimidating than inviting.[22]

Still others saw the car as just another demanding possession. Writing in the 1950s, John Keats, in *The Insolent Chariots*, likens the fashionable changes made to the automobile to the demands made by an aging harpy. "Quickly the automobile became the nagging wife," Keats writes. "Demanding rubbings and shining and gifts. She put eyebrows over her windshield in the 1920s, plucked them out in the late 1930s, put them on again in the middle 1940s, and took them off once more in the 1950s. She nagged him for bits of chrome and cursed him for his extravagance when he brought them home. She lifted her face—expensively—from year to year; incessantly demanded new and different colors, developed ever more costly eating habits, threatened to break the family budget, and often succeeded."[23]

Indeed, the equivalency between the female body and the car body was drawn so early and so clearly that it was caricatured in a May 1920 *Vanity Fair* cartoon. Titled *Sketches of New Body Designs: Drawn for the "Motorist Who Is Contemplating Matrimony,"* the cartoon draws parallels between five new car models and five types of women. "A Stylish Foreign Car," the copy reads, is of "limited production" and is "usually carefully painted and decorated." The Ford is a "non-skid model, that is simple, dependable and designed for all round use," perhaps owing to her "body built with plain, straight lines." Finally, the Limousine, with its "extra heavy, solid-cast body, luxuriously upholstered," is touted as being both "roomy and comfortable."[24]

"Such portraits, although humorous, reduce women to function and appearance, emphasize their status as commodity, and draw attention to physical characteristics," Laura L. Behling writes in "Woman at the Wheel: Marketing Ideal Womanhood, 1915-1934." "That a whole woman can only be understood by mechanically minded men as the sum of her parts, relegates her to the role not of consumer, but to the object, that is, like the automobile, consumed."[25]

Equating women and cars was common in midcentury America. As Beth Bailey writes in *From Front Porch to Back Seat: Courtship in Twentieth-Century America*: "Both were property, both expensive; cars and women came in different styles or models, and both could be judged on performance. The women he escorted, just as the car he drove, publicly defined both a man's taste and his means."[26]

And maybe that's part of the reason why as cars became more comfortable and easier to drive, protective gear was replaced by fashionable "car" clothing. In 1953, aware of women's continued desire to look good in their cars, the Ford Motor Company developed a line of "Motor Mates"—coats and accessories that matched their Victoria model. The handbags were made from the same nylon used in the upholstery and were advertised in *Vogue* as coordinating with both the interior and exterior of Ford cars. Ford dealers were encouraged to sponsor fashion shows, give the coats to local actresses to wear on television, and hold essay contests encouraging customers to write short testimonies about their love for the cars and, perhaps, their matching accessories. All as a way to help women make a place for themselves in the overwhelmingly masculine world of motoring.

The link between fashion and cars would continue for decades. Chrysler, for its part, outfitted the 1955 La Femme with specialty tapestry featuring pink rosebuds on a pale silver-pink background with pink vinyl trim. The car came with a matching keystone-shaped pink calfskin purse with a coordinated compact, lipstick, and cigarette case tucked inside. The purse was made to fit into the back of the passenger seat in such a way that the owner's initials—engraved into the front of the purse—faced out. Also encased into the back of the driver's seat was a compartment that contained a raincoat, bonnet, and umbrella made to match the rosebud interior, all included in a special option package that cost $143.

In 1952, to celebrate the 100th anniversary of the company, Cadillac commissioned jewelers such as Van Cleef & Arpels and Harry Winston to replicate its ornate crest in jewels to be worn by one lucky lady.

"In 1960, the advertising program hit its zenith in design and proportion," one car critic wrote. "A rich blue Sixty Special is matched with diamonds and sapphires crafted by Black, Starr & Gorham. A bright red Sedan de Ville is graced by a crest of diamonds and rubies by Harry Winston, Inc. Van Cleef & Arpels returns with diamonds and emeralds matched with a luxurious deep green Sixty Special convertible. Finally, Cartier's crest of diamonds in their platinum setting crowns two ads; a yellow Sedan de Ville and a white Sixty Special staged in deep blue-violet surroundings."[27]

But the combination of fashion and car design did not always go well. In the 1950s Ford lined their seats with a stretchable plastic that spread when you sat down and reverted to its original shape when the driver or passenger arose. The problem was a big one for women with ample buttocks who eschewed a cloth coat in favor of a fine sable one.

"It turns out that if a woman wearing mink or sable sat on the cloth, the guard hair of the furs would penetrate the interstices opened in the stretched cloth, and then when the lady's weight was removed, the cloth would spring back faster than the guard hair could be withdrawn, and the furs would be trapped," explains John Keats in *The Insolent Chariots*. Mink and sable coats were quickly destroyed, and a process known as "minking" was soon applied to all fabrics used in the automobile interior.

Both cars and fashion demonstrably changed in the late 1960s, just as Baby Boomers were reaching car-buying age. Ford research from the period indicated that buyers were becoming more educated, more sophisticated, and more willing to spend cash for what they called "image extension."[28]

Among the most sought-after vehicles for this young cohort was the Mustang: a sporty yet practical car with an attractive silhouette at an attractive price. It was the car that made Lee Iacocca's career. As a young executive he had championed the Mustang, changing the automobile industry forever. A charismatic, even iconic, automobile executive, Iacocca went on to lead Chrysler as it dipped in and out of bankruptcy. Later he would launch an unsuccessful bid for president and then successfully raise millions to renovate the Statue of Liberty.

While there is some evidence that the Mustang was named after a World War II fighter plane, the official Ford press release from 1962 states unequivocally that the car was named after the horse. "The Mustang is aptly named: Mustang horses are small, hardy, and half-wild. The diminutive two-seater that just trotted out of the Ford stable fits the description."[29]

The first Mustangs to roll off the production line were in the patriotic colors of red, white, and blue, and the Mustang was formally introduced to the public in 1964 in the Ford Pavilion at the New York World's Fair, where celebrities such as Walt Disney and Dr. Martin Luther King Jr. took a spin in the newly introduced model.

Shipped under cover of night to dealers around the country, the new "pony car" drew an estimated four million people to showrooms the weekend it was unveiled, and both *Time* and *Newsweek* had photos of the car on their covers. It took less than two years for Ford to sell one million Mustangs.

Prime among these buyers were women and young adults. Ford spent $10 million on ads in twenty-four national magazines to announce the car's arrival. Or as *Newsweek* put it: "Ford is spending $10 million to embed the Mustang in the national consciousness like a gumdrop in a four-year-old's cheek."[30]

Advertisements, says historian Gary Cross, "are scripts of social dramas that helped people cope with modern life by giving goods meaning and making them into props that said who consumers were or aspired to be."[31]

The Mustang famously promised adventure to office girls, and television commercials featured secretaries embarking on real or imagined adventures with the tagline: "Mustang, everything you could ask for on a secretary's salary." In the ad the bespectacled woman, her hair tied back in a bun, looks inside the car to see her wondrous future in it, "including a tuxedo clad man with roses, a trip to Paris with two handsome men, winning a beauty contest and also getting married!"[32]

Women fell hook, line, and sinker for the Mustang, with its tagline: "More smiles per mile." Perhaps that's why the Mustang, with its upbeat vibe, was famously television star Mary Tyler Moore's car of choice. Modestly priced at just under $3,000, it was engineered to create excitement among an ever-expanding set of Baby Boomers, including women.

In the popular television show, named after the plucky and pleasant actress, Moore works at a local television station in Minneapolis after dumping her commitment phobic boyfriend and in the show-opener she drives her white Mustang into town while the theme song entones: "You're gonna make it after all." It was an easy and early feminist lesson set to a catchy tune and one I remember well.

Mary buys a different Mustang in season three after bemoaning the price, complaining that car salesmen were "pushy" and charming the service department with her well-put-together outfits, causing one service man to remark, "She takes care of her car, the way she takes care of herself" and another mechanic to show up at her office and try to date her.

But perhaps the most egregious Mustang ad is one that ran in 1967 that encouraged women to take the "Mustang Pledge," a pledge that included, among other things, sticking to her diet and promising to "keep the 'helpless female' look by shifting manually only when driving alone." When in the company of men, so to appear unable to work such powerful machinery, women drivers were to pledge to resort to an automatic transmission option called a "SelectShift."[33]

There are a few, but not too many, great motoring stories that involve women out for a ride together. There is the duo of Gertrude Stein and Alice Toklas, who after learning to drive from Parisian taxi drivers, head to the front to drive ambulances in World War I. Reportedly, they never master driving in reverse, a fact that leads Toklas to later quip that they were like the French army—they never retreated.

Then there is the Black writer Zora Neale Hurston and the White writer Fanny Hurst. Theirs is a trip that begins with the invitation to jump in Hurston's beloved Nash automobile, a car that Hurston names Sassy Susie: "Come on, Zora, with your car and let's you and I go on a trip."[34]

Then there's the famous pair of Alice Roosevelt and Countess Marguerite Cassini, whose unchaperoned exploits in cars with both men and women were so scandalous that President Theodore Roosevelt reportedly remarked: "I can either manage the country or I can manage Alice."

Perhaps that is what my sister and I fancied ourselves—two fashionable women out in her little red Mustang. And it did feel that way to me when I was younger. Yet as an adult I must admit that the ride I remember most clearly with my sister was anything but a pleasure ride.

❋

It was my dark-haired sister who had come to get my mother and me after she got the call. It was just after 9:00 P.M. on a weekday. It was hot. A light summer rain had hit the streets just hours before, and steam was rising from the pavement.

We piled into her little red Mustang and headed to the only hospital in town, the one I had been born in. At nine years old, I was still too young to visit a regular hospital room, let alone the emergency room, so I was left alone in the waiting room.

Bright lights beamed down on me. I sat on a small couch and ran my hands over the tears in the upholstery. Empty Styrofoam cups surfed on Formica tables, and old newspapers—their coupons torn out—lay crumpled on the floor.

My father had been driving a Dodge Charger, the epitome of an American muscle car: fast and loud. He used to sell them to boys back from Vietnam, wallets flush with battle pay, the fear and allure of death still close at hand. On summer nights you could hear their tires peel off the asphalt as they screeched around corners, laying down rubber in the hopes of impressing some girl or letting everyone know how pissed off they were about something, or maybe, nothing at all. They were shamelessly macho.

But on the night of his accident, my father's high-powered performance automobile did him no good. He was stopped at a red light when he was hit head-on.

The accident would leave him with a permanent traumatic brain injury, a diagnosis that did not exist then. All we knew was that he had a gash that ran all the way from the top of his head down to his eyebrows. The extent of his unraveling would reveal itself only later.

For all the chaos that had ensued before the accident, everything after that was icing on the cake.

Before the accident, my father was just a garden variety kind of drunk. But as a kid you never understand that. You think he's the worst, the most awful, the most embarrassing ever. You think if you are perfect enough, no one will notice him. That somehow you will be magnificent enough to distract them. But, of course, everyone knew. Everyone always knew.

At home he would sit around in his white undershirts, the kind that used to be called wife-beaters. He drank Budweiser out of quart bottles. When he finished one, he would drop it under the coffee table and say inexplicably: "One little dead Indian." By the end of the night, it was a massacre under the coffee table.

After the accident, he became a violent raging drunk. But it was nonsensical raging. Brain-injury-induced or alcohol-induced raging that wouldn't fit into any clear pattern. There was no logical reason for any explosion on any given day. Like the boys out squealing their tires, Dad's raging was more about wreaking havoc than putting any kind of fine point on an argument.

The stories became darker, the kind of thing you couldn't even whisper about to friends. I remember one day he came home and beat the dog nearly to death while I cowered in the kitchen and listened to the yelping. There was no rhyme or reason to it. Whether it was brain-injury-induced or alcohol-induced—I never knew.

He attempted to go back to work but had lost the sweet-talking patois of sales and the ability to run the manipulative processes necessary to close the deal, and our income eroded swiftly. Events on the world stage were also rocking our world in Waukegan.

By the 1970s, as I was about to enter high school, sales of new cars began to fall precipitously as OPEC (Organization of the Petroleum Exporting Countries) tightened production and gas lines formed at the pump. President Jimmy Carter would famously appear on television in a cardigan warning us of the dangers of dependence on foreign oil and looking like Mister Rogers in the Oval Office.

"Suddenly," according to Jeff Rothfeder in *Driving Honda*, "the ideal car had a new set of criteria: instead of fancy, engorged designs, big chassis, and powerful engines, people sought high gas mileage and motors that ran efficiently and inexpensively for much longer than those of the typical American cars of that period."[35]

The production of high-speed muscle cars would drop precipitously in the ensuing years, and, as a result, Detroit would lose its single-handed grasp on the automobile business. In 1974, Honda introduced its four-door Civic, which met stringent Clean Air Act emission standards that called for a 90 percent decrease in carbon monoxide, hydrocarbon, and nitrogen oxide levels.[36] It was the first car to meet those standards and was introduced at a time when Detroit's big three automakers were arguing vehemently that it couldn't be done. The slow unravelling of the American auto industry had begun. Chrysler would teeter on the brink of bankruptcy and then emerge again. But neither my father's business nor his brain would ever fully recover.

FOUR

The Moving Van

I live my life a quarter of a mile at a time.
—Dom Toretto, *The Fast and the Furious*

Deviled eggs, three-bean salad, and a trip to the racetrack every weekend. Those are the things I remember about growing up in the Midwest. My brother drove a race car at the Speedway up north in Wisconsin on weekends to the sounds of shrieking girls and the ever-present smell of burning rubber.

After the accident, my father's house—where my siblings lived and I visited on weekends after my mother and I moved back to our bungalow—became more chaotic than ever. He was unable to mow the lawn or keep up with basic maintenance, so the grass grew knee-high and screen doors flapped wildly as the wind whipped off Lake Michigan.

We lived in an area of small but tidy homes, but our lawn, such as it was, was scattered with broken cars of different makes and models in need of various repairs or parts. Chevys sat on blocks beside a Mustang missing a door and a stock car covered with grease.

Maybe because of the cold winters or because of the unreasonably high winds that swept in off the lake, my father and brother also kept spare parts in the house. Tires, chains, mufflers, and oil pans—all piled high next to the couch, on the television, and on the dining room table. Some of the parts were for regular cars, but many were specifically designed for my brother's dragster, which he took to the track on weekends on a trailer bed attached to his pickup.

My brother was thirteen years older than I. His name was Vanderbilt, but everyone called him Van. He was tall and thin, with greasy hair and ears that stuck out from the sides of his head. In winter, when he couldn't race his car at the Speedway, he played pond hockey behind our house, whipping pucks into a makeshift net with breathtaking speed.

One summer when I was eight or nine years old, wanting to play catch with him, I went running after a baseball he had tossed sky-high. Despite his warnings, I swooped in to catch the ball barehanded. The ball slapping into my hand produced a pain I still remember today, one that spoiled my desire for almost any interaction with him, most of which seemed to hurt in one way or another.

A few months before my father's accident, we'd all gathered at a local tavern called Louie's to await my brother's return from his draft physical. No longer married but linked permanently by parentage, my mom and dad held hands under the table as they waited. Towns like Waukegan supplied a steady stream of boys for the Vietnam War. They were summoned to a large armory in Chicago, where they followed red, yellow, or blue lines for physical and mental testing that would determine their futures.

Never a picture of health, my brother was six foot three but had a kind of scoliosis that made him appear slouched even when he was standing up straight. He also had ferocious acne that would last his entire adult life. His general ill health made him an unlikely recruit. Half a dozen of his friends had already been called up—boys without a hope of deferment based on college acceptance or a letter from a well-placed relative.

My brother lurched through the door and flashed a goofy smile and a quick thumbs-up. His category was 4-F, which translated roughly at the time to "forever free." My mother erupted in tears of relief, and my father gasped then awkwardly rose to hug him. "Atta boy, Butch," he said. "Atta boy."

My brother called his drag car the Moving Van. On weekends, he towed it to the track, where he drank beer and ogled women who wore halter tops and hoop earrings. Even in the wake of the car crash that almost killed our father, my brother would continue to race every weekend.

If the advertising and marketing of the car is all smoke and mirrors, the car itself has an irresistible and precise logic: the wheels must be aligned, the fuel must be injected in the proper amount at the proper time, and the pistons must fire in an even rhythm. "The car amounts to a kind of logos, a self-contained system of causes and effects, a wholeness of truth and reason," according to Daniel Guillory in "Bel Air: The Automobile as Art Object."[1]

If this is true in general about cars, everything is exaggerated when you race. In drag racing, everything must go correctly within a very short amount of time to achieve the kind of acceleration needed to go fast off the line. Drag racing, therefore, requires a rhythmic acceleration, climax, and deceleration. Beginning, middle, and end must all be properly proportioned. Done well, it can be exhilarating. Done poorly, it can be dangerous, even disastrous.

As a young girl, I would watch my brother as he got the car ready and listen to the banter of the men around the car. Later I would slide into the stands, watching as his car got towed or pushed to the line. Sometimes he won; other times he would grind through his gears too quickly, putting too much pressure on his axles, causing them to break, sending the car into a smoky slide. He would end such rides climbing out of the car window, whooping and hollering and high-fiving, as happy as if he had won. In a way he had; once again he had defied death and triumphed over that particularly male domain of speed, time, and space that the drag strip epitomized.

*

As the drag strip demonstrates, the invention of the automobile compressed time and space, allowing us to move through and beyond geographies in heretofore unknown ways. And while the drag strip was one particularly male example of that compression—the very fact that we could be someplace else in a short amount of time would soon lead us to the state where we felt we needed to be someplace else in a very short time. More options would lead to more encounters, and a more fractured sense of time and space. Since we could now do more, we suddenly felt that we should do more. In old German—the country where the car was born—the word for *speed* and *progress* share the same root verb.[2] Progress or forward motion—often in a car—at great speed would become the watchword of the twentieth century. Women, however, would have to proceed with caution.

Women, after all, were never quite handed *le droit à la ville*, the right to walk aimlessly within the walls of the old city without being challenged, threatened, or hassled. As Lauren Elkin points out in her book *Flâneuse: Women Walk the City in Paris, New York, Tokyo, Venice, and London*, historically a woman on the streets was most likely a prostitute and even her movements were strictly regulated.[3] Women started to appear in public more frequently with the advent of the bicycle, but it was the car that gave them prominence in terms of both visibility and power. "Women never explicitly demanded the Droit de la Ville, the medieval right to the freedom of the city that distinguished urban citizens from feudal serfs," writes Virginia Scharff in *Taking the Wheel: Woman and the Coming of the Motor Age*, but "the auto had the potential to help some women claim that right, most without ever demanding it."[4]

For that reason, when women began to embrace the automobile—taking to the streets for pleasure or to meet work, family, and farm obligations—their reception was frosty indeed. Wealthy women went first, protected by their class status, followed by rural women, who fought isolation and destitution and quite simply needed to get stuff done.

Women drivers, it was safe to say, were met with resistance at almost every turn. Fear of women drivers, however, was less about concern over car accidents and more about discouraging a behavior that was widely threatening not just to men, but to the larger social structure in general. "From the beginning," historian Michael Berger writes, "everything about the car seemed masculine, from the coordination and strength required to operate it, to the dirt and grease connected with its maintenance."[5]

That is likely why women chose the car as the proper symbol of their independence. For example, in their quest for the right to vote, female suffragists completed a widely publicized trip across the United States in 1915 in an Overland car provided by the company. "The sight of the car, with its worn banners, luggage roped on to the running boards, and the western mud and dust carefully preserved to heighten its effect, allowed the suffragists to stage a highly stylized performance of female independence, transcontinental solidarity, and sexual equality," historian Georgine Clarsen states in *Eat My Dust: Early Women Motorists*. "It was a nonverbal vocabulary of entitlement and social change, perfectly intelligible and easily read by people in the streets."[6] The Overland Automobile Company was quick to capitalize on the women's achievement, feeding to local newspapers details of the women's trips and specifics about the car they drove and what they wore, setting in motion the beginnings of a long line of automobile promotions aimed at women, and linking forever social striving with the right car.

Yet despite their extraordinary efforts, women drivers were often the butt of jokes and downright intimidation. Men thought women too weak, too emotionally unstable, to be on the road. The fear of women drivers, like most myths, probably also had some basis in fact. Not because women lacked skill, competence, or emotional stability, but more because early machinery was difficult to operate—for everyone. Early cars were hard to crank and even harder to steer. Likely most readers of this book have never driven a car without power steering and enhanced suspension, and it is worth remembering here that the super-smooth superhighways that we have come to know and expect are a post–World War II phenomenon, one that is a far cry from the rutted, muddy roads that early women drivers had to navigate.

Almost all histories of motoring locate the introduction of the electric starter motor as the turning point for women motorists. According to an account in the *Cadillac Craftsman*, a trade paper produced by the company, Cadillac president Henry Leland asked electrical engineer Charles Kettering to produce an electric starter in 1910. Leland was reportedly inspired to do so after losing a good friend who had died after gallantly cranking a car for a woman on Belle Isle Bridge in Detroit. In the official version published in the *Cadillac Craftsman*, the "strange" woman had omitted to retard the spark, so that when the gallant gentleman and friend of the Cadillac president spun the crank, the handle kicked back and broke his jaw. He later died of gangrene.[7]

The electric starter would come to be called "the ladies' aide." And the story of the gallant gentleman would become part and parcel of automotive history, along with the idea that women lacked either confidence or competence or both when it came to cars—a baked-in gender-based inferiority that could be neutralized only by technological advance.

Nevertheless, as Clarsen tells us: "One could easily theorize that Charles Franklin Kettering did more to emancipate women than all the feminists of the day. Before his invention, women could drive if the engine were running. But the lusty Amazons who could crank one were very few indeed. Then in one broad stroke, the dainty female toe became as powerful as the brawny male arm."[8] And that changed everything.

✳

Women, who fought at every turn for the right to drive, were most definitely not given the right to speed. Racing was by and large the privilege of wealthy men. Early on, internal combustion engines and the massive cars they propelled appealed to powerful men like William K. Vanderbilt as a status symbol and a marker of manhood. "They used large, powerful gasoline-fueled automobiles to set themselves apart and to communicate the message that they were privileged men," according to Tom McCarthy in *Auto Mania: Cars, Consumers, and the Environment*. "That it took money, skill, and 'balls' to drive an automobile."[9]

The car would do more than change the way we got around—it would become part and parcel of how we defined ourselves, particularly our gender. Manhood had long been defined simply as the opposite of childhood, but around the beginning of the twentieth century, *manhood* was replaced gradually by the term *masculinity*.[10] And one powerful way to express your masculinity was by driving fast and making a lot of noise. Speed and aggression have always been the province of males, especially when it comes to cars. But with the advent of Charles Kettering's starter, aka "the ladies' aid," and the power of the internal combustion engine, technology began to change how we lived and how we viewed ourselves. "The car's overt role in making modern men has obvious implications for making modern women," Deborah Clark concludes in *Driving Women: Fiction and Automobile Culture in Twentieth-Century America*. "In challenging male automotive dominance, women challenged masculinity itself."[11]

As a result, women were generally discouraged, even prohibited in some cases, from driving powerful vehicles with great range and speed. That's why Clara Ford, the wife of Henry Ford, drove a 1914 Detroit Electric—not a Ford. In the first instance, women were discouraged from driving gas-powered cars because of safety concerns. Bicycle and electric vehicle maker Albert Pope famously derided the gasoline-powered automobile, arguing that "you can't get people to sit over an explosion."[12] According to one expert, "In 1900 in Chicago, for example, no licenses were given out to women for steam and gasoline cars, because these were considered 'unsuited for use in feminine hands.'"[13]

Women were also discouraged from driving gas-powered automobiles for a second, more prurient reason. Social critics believed that the vibrations from the motor would cause sexual arousal, a concern first raised as an argument against the bicycle and one that carried over powerfully to the remarkably shaky early car. Electric vehicles had one other advantage: they had a limited range of around fifty miles and lacked enough power to manage large hills.[14] Clara's vehicle topped out at twenty miles per hour, pretty much enough to get her around her enormous estate, Fair Lane, and make a social call or two.

Electric vehicles therefore allowed for necessary, routine errands, social visits, and the occasional appearance in a parade, but more or less prohibited a woman from gallivanting about the countryside unchaperoned or using the car to escape her domestic duties for long periods of time. Still, the electric car was the choice for early women motorists who could afford them. Wealthy women such as Queen Alexandra and the Duchess of Alva drove or owned electrics. Queen Alexandra reportedly used hers to tour her estate in Sandringham, Norfolk. Her Majesty was reportedly "delighted with the ease and simplicity of control."[15] Stateside in a *concours d' elégance* in Newport, Rhode Island, in 1899, socialites such as Mrs. Ava Astor drove their electric cars festooned with flowers in one of the largest automobile parades to date.[16]

Yet there were a handful of exceptional women who eschewed the electric, preferring speed, danger, and multiple sexual partners. Hellé Nice, the daughter of a French postman, raced in the 1920s and 1930s all over Europe, sponsored by the luxury carmaker Bugatti, leading her biographer, Miranda Seymour, to dub her "the Bugatti Queen." The former stripper was the lover of Philippe de Rothschild, who lent her his own Bugatti and then introduced her to famed carmaker Ettore Bugatti, who sponsored her racing efforts in a bid to get wealthy women interested in his cars. After winning the 1929 Grand Prix Féminin, she stopped for a moment to fix her makeup, pierce the blisters that had formed on her hands during the race, and change into an elegant gown, before telling journalists with much sexual innuendo that the pleasure she most adored was "the feeling of a great engine roaring and under your control."[17] After Nice set an early land speed record, the French sporting newspaper *L'Auto* called her "une conductrice de Valeur," or a very fine driver, and then added, *for a woman.*[18]

Nice would later be involved in a horrific accident in São Paulo that killed four and injured thirty-six spectators. Toward the end of her career, Nice raced in an Alfa Romeo designed for her by one of her many lovers—it was a shocking turn of events for the auto press that a French woman could triumph in an Italian car. In the wake of World War II, a male race car driver's claims that Nice had been a Nazi sympathizer would end her career, and Nice would die penniless, reportedly stooping to stealing milk out of a cat's saucer to survive.

Florida native Betty Skelton, who first made her name as an aerobatic pilot, set a stock car record in a 1954 Dodge Red Ram V8. She went on to set the women's land speed record with a 430 kph (276 mph) average. Skelton earned four female world land speed records and set a transcontinental speed record.

Another favorite is Jacqueline Evans de Lopez, who raced in the 1960s in the rugged Carrera Panamericana—a border-to-border race in Mexico—in her Porsche adorned with the likeness of Eva Perón.

But the woman I remember most was Shirley "Cha Cha" Muldowney. Muldowney sometimes raced at the track my brother frequented, and they called her the First Lady of Drag Racing. I remember watching her get towed to the line in her pink dragster. Watching Muldowney race was one of my first lessons in feminism. While most of the macho culture and noise of racing was intimidating to me, there amongst the smoke and the dirt, was a petite woman. Even if I didn't understand what feminism was, it was hard to miss her appeal. I kept a small picture she signed for me on the wall of my bedroom.

While most of the women in the stands boasted hot pants, halter tops, and hoop earrings, Shirley—my girlhood idol—was playing a man's game and winning. All told, she would win eighteen national titles. For young girls growing up in the Midwest in the 1960s and 1970s, Shirley was likely the first woman we ever saw break a gender stereotype. Moreover, she looked like she was having fun while doing it.

Still, as I watched the smoke clear on a hot summer's evening at the track, it was clear to me that this was a world I was excluded from; I knew that I would not be part of the culture that defined my family. My femininity, my womanhood, my future was defined at that moment not by what was possible, but by what was prohibited.

FIVE

My First Car: Speed, Power, and Women's Liberation

The charm of automobiling isn't all in whirling over miles of dusty roads in a mad race against time. There's something in the personality of the motor, and half the happiness consists in being convinced, in your own soul, that yours is "The Very Dearest Car in the World."

—Hilda Ward, *The Girl and the Motor*

I was ten years old when my mother died; I was numb and fragile. The days after her death were a haze. Yet one thing I remember clearly from that time was the black Cadillac in the driveway, purring as it waited to take us on that long smooth ride to the funeral home. I remember the warm leather seats and the Windex smell from freshly cleaned windows and the small box of tissues that rested at my feet on plush, newly vacuumed carpets.

Soon after my mother's death, my sister married, and my brother left home. I was alone with my father, who rallied—until he didn't. There were vast swaths of uncertainty, each of us lost in our own sadness and disassociation. At home reality wasn't ever one set thing but an ever-changing array of choices. Like wheel trim and interior options, you could take your pick. Dinner would be chicken at five, or noodles at seven, or maybe there was never a plan for dinner in the first place. There would be a birthday party for you, only there wasn't. He was going to pick you up from school, but he never showed. It was cold, sometimes below zero, and the snow was blowing. You waited until you couldn't stand it anymore and eventually started walking.

Our financial footing was unstable. My father worked most days from 10:00 A.M. to 9:00 P.M., with Thursdays and Sundays off. He still sold cars, sometimes used and sometimes new, with varying degrees of success; but once the gas crisis of the 1970s hit home, the muscle car began to lose its appeal in favor of more fuel-efficient Japanese cars, or "souped-up lawn-mowers" as my dad called them.

Sales were sometimes good, sometimes not. And you never knew one way or the other until the bill collector called, or, worse, came to the door. We were finally evicted one cold winter's night.

In the wake of the accident, my father's steady stream of lies had become a torrent. "When we discover that someone we trusted can no longer be trusted," the poet Adrienne Rich once wrote, "it forces us to reexamine the universe."[1] My father's lying caused me to doubt everything. I once got a D in geography from a teacher who felt I was being impudent for asking her how sure she was that Africa actually existed. "Have you ever been there?" I asked. I thought it a reasonable, even urgent, question. She sent me to the principal. She thought I was just being cheeky, maybe even rude.

Still, I wanted to believe, so I told myself I was the one who was confused. I was the one who got it all wrong. And it went on like this for a long time. Until I doubted not only myself, but reality itself. Reality became as slippery as midwestern winters.

My report cards from those years tell the story. Pleasant and compliant when engaged, I would as often as not simply drift away when not being spoken to directly. The state of Illinois required standardized testing, and I remember having to go to the principal's office to retake the tests orally. On test day, I had sat in my seat without filling in even a single bubble on the test sheet. I hadn't flunked, exactly, but I had been completely absent while being fully present. "Being traumatized is not just an issue of being stuck in the past," writes Bessel van der Kolk in *The Body Keeps the Score: Brain, Mind, and Body in the Healing of Trauma.* "It is just as much a problem of not being fully alive in the present."[2]

After her death, my mother's old blue Chevy Impala rusted away in the front yard, and I would sometimes sit in it, running my hands over its seats and sniffing the interior to see if it held any remnants of her perfume.

Books became my anchor. The words were printed in black and white, steady and unmovable, there to be checked and rechecked again, their logic forever bound together with a simple seam.

At home, we had only a half-complete set of encyclopedias that my father had bought from a college student going door-to-door. I suspect he bought them more out of some kind of salesman simpatico than a lust for knowledge, and he must have arranged a payment plan that was never completed, since our set stopped with the letter *M.*

We didn't get any newspapers delivered to our home, and we didn't get any magazines. Instead, I relied on Gertie. She was the town's bookmobile. The name came from the manufacturer, the Ohio-based Gerstenslager Company, which had originally made horse buggies but later specialized in retrofitting buses for multiple uses. Most famously, the company made five Wienermobiles, vehicles in the shape of hot dogs, for the Oscar Mayer Company, and later a vehicle in the shape of a Cubs baseball hat that I've seen in pictures but don't recall ever seeing in real life.

In summers, to help fill the long, empty hours of school-age children, Gertie moved from neighborhood to neighborhood, supplying books. Librarians, being no fools, were smart enough to park Gertie next to the candy store or the town pool. On hot summer days, Gertie had the remark-able advantage of being air-conditioned. She offered salvation, two books at a time.

※

After books, however, my car was the first thing that ever linked me to anything else. Finally, I could control my own schedule, keep a part-time job, visit friends when home became too violent or just too damn loud and unpredictable. Is it an overstatement to say I didn't come alive again until I could drive? Probably, but that's how I remember it.

Driving forces you into the present. Will that car in front of me stop? Is that other car turning? I simply had to pay attention.

When I turned sixteen, in 1975, my father swapped an old snowmobile and an old boat for the kind of car commonly known as a "beater." It was an enormous old Chrysler Newport, two-toned gold and brown with a tan interior, beige in both color and affect, and enormous. My father said he got it for a song.

I learned to drive in high school, avoiding the pitfall of many young female driving students, the roving hands of the male instructor. Mine was well-behaved, although stories abound of young women being assaulted by instructors.

It was a relief to drive by myself in a car for the first time. I remember vividly getting behind the wheel for my first solo drive. I got in and pulled the door shut. It made a kind of clunking sound as it closed. I put the car in reverse, and it made a popping sound like the joints of an old man. Cars today are largely made of plastic, but back then they were all steel and their sounds were distinctive—they signaled that things were all right, or going to be all right, or at least could possibly be all right in the near term. The car had gas, the weather was good, the brakes would hold, and it was time to go.

I backed out of the driveway and onto the hardtop of the road. A light rain had done little more than darken the pavement. It was there that I felt the car accelerate as I gave the pedal a little push up the hill. My car was ugly, but it was fast, with a 360-cubic-inch V8 engine. The feeling was unmistakable: pure power and freedom. Like most cars back in the days before four-wheel drive and antilock brakes were common, mine was a sled in the snow. I used to take it out into the country to turn doughnuts. Putting a foot on the accelerator and a foot on the brake at the same time made it turn wide, lazy circles on icy pavement and snow-covered lots.

In summer, we would take a six pack of beer down to the lakefront. One person would drive, and another would hook her feet under the dash and hang out the passenger-side door, dragging a can along the pavement to create long trails of sparks behind us.

"Cars magnify bodily powers, so that smaller and finer bodily move-ments are required to release and control progressively larger amounts of automotive power," according to Georgine Clarsen in "The 'Dainty Female Toe' and the 'Brawny Male Arm': Conceptions of Body and Power in Automobile Technology."[3] Driving freed me from the neglect, boredom, and abuse I found at home. It changed not only the parameters of my life—it changed my experience of life. No longer left to my fantasies and my imaginary playmates, or the dogs we couldn't afford to feed, or the random spare parts in the front yard, I could indulge in long leisurely moments in my car. I could drive to the library, to the grocery store, and to get an ice cream cone. And make no mistake: life looks different at forty miles per hour.

Indeed, our ability to drive, to see the world from inside the automobile, with its constant ever-changing landscape, has changed the way we both see and experience reality. Long before there was the internet, the fractured, abstract, and blurred landscape we saw from the car became, in a very real sense, how we saw the world. The sense that time could pass so swiftly and so disjointedly allowed us to embrace the widespread visual abstraction that became common after World War II as film and television changed our conception of time and space.

"The actual experience of driving on the freeways prints itself deeply on the conscious mind and unthinking reflexes," Reyner Banham wrote in his 1971 book, *Los Angeles: The Architecture of Four Ecologies*. Driving, especially highway driving, is "a special way of being alive."[4]

Consider the fictional character of Maria, a Corvette-driving small-time actress, who cracks hard-boiled eggs on the steering wheel while shifting smoothly on the interstate, never missing a beat or an exit. Maria is divorcing a director named Carter and generally unspooling in that cool California way that Joan Didion made famous in her novel *Play It as It Lays*:

> In the first hot month of the fall after the summer she left Carter (the summer Carter stopped living in the house in Beverley Hills), Maria drove the freeway. She dressed every morning with a greater sense of purpose than she had felt in some time, a cotton skirt, a jersey, sandals she could kick off when she wanted the touch of the accelerators, and she dressed very fast, running a brush through her hair once or twice and tying it back with a ribbon for it was essential (to pause was to throw herself into unspeakable peril) that she be on the freeway by ten o'clock. Not somewhere on Hollywood Boulevard, not on her to way to the freeway, but actually on the freeway. If she was not, she lost the day's rhythm, its precariously imposed momentum. Once she was on the freeway and had maneuvered her way to a fast lane, she turned on the radio at high volume and she drove. She drove the San Diego to the Harbor, the Harbor up to the Hollywood, the Hollywood to the Golden State, the Santa Monica, the Santa Ana, the Pasadena, the Ventura. She drove it as a riverman runs a river, every day more attuned to its currents, its deception, and just as a riverman feels the pull of the

rapids in the lull between sleeping and waking, so Maria lay at night in the still of Beverly Hills and saw the great signs soar overhead at seventy miles an hour.[5]

In that first summer after she left Carter, Didion notes, Maria put seven thousand miles on the Corvette. It was driving that saved Maria that summer—just as driving saved me.

❋

The car has been called our most gendered technology, and for good reason. Even as the car accelerated women's quest for independence, accentuating both our range of movement and our ability to engage in the larger economic world, it also reinforced earlier, more limited, versions of what it meant to be a woman.

The soft female organic form has always stood out in sharp contrast to modern technological machinery. But, as Maria demonstrates, the modern woman would be constructed, bit by bit, by her use of the modern automobile. No one was more challenged by this than the automakers themselves, who were forced to confront and to manufacture a new woman to both use and purchase their products, and they needed to do so in a way that did not alienate men.

"Women took to the wheel in ever increasing numbers and automobile manufacturers and advertisers were forced to reckon with the tension produced by the recently politically liberated consumer," Laura L. Behling writes in "'The Woman at the Wheel': Marketing Ideal Womanhood, 1915–1934." "However, their strategy was not to redefine woman for the independence she exhibited, but rather to carefully contain her in traditionally feminine sex and gender expectations."[6]

The same year that I received my license from the state of Illinois, for example, another woman, Barbara Roos, was not allowed to get a driver's license in her maiden name in Florida but was instead ordered to use her married name.[7] That was just two years after the passage, in 1974, of the Equal Credit Opportunity Act that allowed women to get car loans. But let's be clear: overwhelmingly, women were expected to use their automobiles in the service of family.

"Belief in a distinctive experience led corporations to key into women as a 'special group,'" Katherine Parkin writes in *Women at the Wheel: A Century of Buying, Driving, and Fixing Cars.* "They tried to persuade America that women needed a car of their own to achieve a circumscribed, domestic freedom and they formulated seemingly countless ways to gender car ownership."[8]

As a result, I used my car not simply for adventure but also to try to keep my brittle little family in place. I got a job to help support us, I did the grocery shopping, and I went to the bank. In sharp contrast to the freedom my brother felt on the racetrack, the car for me was a tool. The car liberated me, but it also made me another worker in the home, someone who could be asked to do way more than was reasonable for a young woman my age.

SIX

The Punch Buggy:
The Volkswagen Beetle
and Violence against Women

*In a sense, the whole twisted ideological mess of 1930s German
politics, with its mismatch between cultivated socialist image
and underlying tendency to violence, is condensed into the air-
cooled, flat-four VW motor.*

—Matthew Crawford, *Why We Drive*

Learning to drive was not only a personal milestone, it was also a
boon for the rest of the family. Now that I could run errands and do
the shopping, I was like every other woman driver. In the broad swath of
automotive history, one thing has proven true: for men the car has always
symbolized adventure and escapism, but for women the car likely means
a longer to-do list. More chores means more miles.

"Women," writes Deborah Clarke in *Driving Women: Fiction and Automobile Culture in the Twentieth Century,* "have paid a considerable price for
the benefits of automobility."[1] As far back as World War I, when women

were suffering wartime shortages and the absence of husbands and sons, they were the target of ads that correlated car ownership with servitude. "In December 1918, the Lexington Motor company ran an ad entitled, 'For Hands that Rocked the Cradle.'"

The ad read:

> Mothers of the boys at the fighting front are giving more than their sons—they are giving themselves, their time, ability, and labor to the at-home work necessary to win the war. For their Red Cross and numerous other activities, a timesaving, energy conserving car, such as Lexington, is of inestimable value to their personal efficiency, and in the aggregate, to their country. [2]

The car in this ad drives home the point. Women were to use the automobile to aid in their service to others. That would also be true in my experience and in the lives of so many women.

So it was no surprise when my brother called me just after I got my license to ask me to pick him up at Chicago O'Hare Airport, about forty-five minutes away from our house.

After my mother died, my brother left home and opened a small travel agency in town. He sometimes went on promotional trips sponsored by the airlines, but in our small town my brother was probably most famous for having left his bride not exactly at the altar, but the morning of the service. My sole memory of that day is waking up to a living room filled with flowers and fancy clothes never to be worn. The aggrieved bride had sent them to our house in a fit of anger, hoping for some well-deserved revenge.

My brother ditched his hometown fiancée, a nurse's aide, for a flight attendant he met on a cruise. I don't remember much about the stewardess, but I remember she was from northern Wisconsin and had enormous breasts—or "headlights" as the boys called them then.

I was just past sixteen by the time the stewardess had landed. The industrial Midwest was losing jobs to nonunion factories in the Sun Belt, and the American auto industry was losing ground to more fuel-efficient Japanese cars.

Despite it all, driving still brought with it the promise of freedom and the allure of power at the tap of the toe. So, at first, I wasn't unhappy to run an errand or drive the toll road forty-five minutes down to O'Hare to collect my brother and his girlfriend from one of their trips.

My car was in the shop, which it always seemed to be. Just like the shoemaker's daughter has no shoes, I was often without a working car. Like an automotive Goldilocks on the lot that our front yard had become, I had several dubious choices for a ride. One car was too fast for me to handle, one too unreliable, one not my style.

The one that was just right was a cute yellow Volkswagen Beetle with a stick shift, an ambitious choice on my part, since I was new to both driving and shifting. This was usually not much of an issue since the Midwest was flat and, back then, the roads were often wide open.

I didn't count on the multilevel parking lot at O'Hare—the one with a great big circular ramp. I made it in all right and went to fetch my brother and the airline stewardess from Appleton. I knew they had been drinking and fighting as soon as I saw them walk off the plane.

We walked together to baggage claim. They argued. We crossed the parking lot toward the garage. They argued some more. We rode the elevator to the top. Floor by floor, they argued. It was only when we got to the car and he saw the little Beetle that my brother stopped mid-fight and said to me, "Jesus Christ, why did you bring that?"

Made by the German carmaker Volkswagen, the Beetle premiered at the Berlin Auto Show in 1939. The Nazi Party had sought to develop a "People's Car" at the behest of the führer, who wanted an affordable form of transport for workers under a program called Strength Through Joy, which included offering cruises, outings, and that version of happiness that can only come on four wheels. The car was part and parcel of the Nazi Party's dream, which was aimed at building a better lifestyle for people, including an easy way to get outside the city to enjoy nature and other leisure pursuits.

"To say that the average German could have 'speed' in his or her control was like a fairy tale," writes Andrea Hiott in *Thinking Small: The Long, Strange Trip of the Volkswagen Beetle*. "But Hitler knew all of this, and like the color red, the swastika, and the democratic vote, he would adopt the idea of a People's Car and twist it into something toward his own ends."

Adolf Hitler turned to his close confidant, the legendary designer Ferdinand Porsche, to design his People's Car for the masses. He must have thought that the "people" didn't have luggage. Or maybe he thought they would take only very short trips with a small picnic basket. The trunk—inconceivably in the front of this crazy, but cute, little car—barely held my brother's bags.

My brother must have known he was too drunk to drive, or he would never have let me take the wheel. Even today, I instinctively give up the driver's side to a man, as do most women. According to a 2010 study, even in feminist households, the man is three times as likely as the woman to drive when a mixed-gender couple is in the car.[3] And there was no question that the men in my household were not in favor of bra-burning women "libbers," as they called the few feminists in town.

In this case, my brother was just too drunk to drive and too intent on pummeling his girlfriend, so he and the blonde squeezed into the car, where they almost instantly began to sock each other. The fight had a kind of comedic element to it, even if it was abhorrently violent. He was in the front, and she was in the back; they often missed, but sometimes they delivered a wallop. He slugged her; she slapped him back in a rhythmic, violent dance. I was relieved there were no tools in the back seat when I spied the down ramp of the garage, all six floors of it.

I maneuvered the car down the ramp slowly. The car lurched forward then stalled in an awkward way, laterally across the ramp, perilously close to the railing. I would get it going again only to release the clutch too fast and stall out again. My brother was barking orders at me the whole time, which only made it worse. And on and on it went: clutch, punch, stall, clutch, punch, stall.

As we hit the toll road, they finally passed out, exhausted from the heat, the alcohol, the nauseating bouncing of my bad gear-handling, and the exertion that it took to beat the crap out of each other. Her blonde hair was plastered against her face, and my brother's eye was beginning to swell. I slipped the car into gear, turned the radio on low volume, and headed for home.

✳

Sometimes called a whore of a car since it was routinely worked on, customized, and then passed around by gearheads, the Volkswagen Beetle has a unique place in car history.

Hitler's first specs for the car were so ambitious they reportedly stunned his master engineer, Ferdinand Porsche. They included a cruising speed of sixty miles per hour, an air-cooled engine, and room for five people. Plus, the price had to be less than one thousand Reichsmarks retail. [4] "The air-cooled VW motor is commonly regarded as 'simple,' but outward appearances are a bit deceptive," writes Mathew Crawford in *Why We Drive: Toward a Philosophy of the Open Road*. Instead, Crawford calls the Beetle "a marvel of precision manufacturing." [5]

Hitler was reportedly inspired to build the People's Car after reading a biography of Henry Ford in which Ford argued that the car would create both additional free time for workers and an enhanced experience during that time. "Unless we know more about machines and their use," Ford wrote, "we cannot have the time to enjoy the birds, the flowers and the green fields."

Hitler must have been convinced since the vision that the two had for democratized, accessible leisure in small, efficient cars was clearly similar, although their strategies differed greatly. Ford, a die-hard capitalist, had reduced the cost of cars by standardizing production and increasing wages for workers to an unheard-of $5 a day.

The Nazis had their own methods. The Beetle "was a socialist car," according to Crawford. "But it was also a fascist car, in the precise economic meaning of fascism: its low cost was due to dictatorial power, state-directed investments, and the outlawing of independent labor organizations not controlled by the Party." As Crawford and others have reported, the manufacturing of the People's Car relied heavily on slave labor, largely Czechs and other Slavic peoples, at Volkswagen during the war.

Production of the Beetle stalled during the war as the great factory in Wolfsburg, Germany, was repurposed for wartime manufacturing. After the war, Wolfsburg, under control of the Allies—mostly in this case the British—restored both the plant and the town where the Beetle was born. The People's Car would sell nearly a million units in the years following the war, first to British servicemen stationed in Germany during the massive rebuilding required by the Marshall Plan, and later to German citizens.

According to one report, the millionth Volkswagen Beetle was painted gold and adorned with shimmering jewels. "Studded with rhinestones," one observer remarked. "Like a tart ready to walk the German streets."[6]

Remarkably, the little German tart of a car would become a hit in the United States about a decade later, propelled by Jewish admen from New York, who sold the cars to peace-loving hippies in the 1960s.

Set at a time when one of the best-selling books of the year was titled *The Magic of Thinking Big*, the advertising team of Doyle Dane Bernbach came up with its own straightforward pitch for the diminutive car. Under the tagline THINK SMALL, the New York ad team emphasized the highlights of the car, pitching its thirty-two miles to the gallon, its aluminum air-cooled rear engine that went seventy miles per hour, and its sensible size and price tag. In other words, it went with the basics. They weren't selling a story about the car, or who the car might allow you to become, they were selling a very real object with specific characteristics.[7]

The ad became one of the most iconic automotive advertising pieces of all time, not to mention one of the great plot twists in automotive history. The Beetle, a car that was born amidst the propaganda machine and destruction of the Nazis, was sold in a straightforward ad campaign created by Jewish admen that focused on performance and whose copy began with the line "This is an honest car."[8]

College kids tacked the ad to the dorm wall, and quickly the affordable little bug became a hit with the younger generation. According to Hiott, the Beetle was just a modern version of the Model T because "it helped open car ownership to people who might not have been able to afford a new car otherwise."

Just a year after the famous ad, the small car represented 46 percent of all car imports to the United States. Indeed, the Beetle would come to define a whole generation of kids who saved up to buy it by working summer jobs and then took it on road trips in cars emblazoned with peace signs and covered in flowers.

But by the 1970s, as OPEC limited oil production and gas shortages racked the United States, consumers started to favor Japanese imports for their efficiency. Beetles officially stopped being sold in the United States in 1978, but Volkswagen made another run at the US market with its redesigned New Beetle in 1998. This time, the focus was on women. And this time, the flowers were on the inside.

Termed a *blumenvasen* (bud vase), the clear acrylic cylinder was attached to the front dashboard. According to Volkswagen, the vase first appeared as an accessory in the 1950s but could be tracked much further back to when flower vases were, according to the company, used to offset the smells that could come from exhaust.

"Automobile vases first started appearing in the late 1800s not as an interesting novelty, but out of necessity," the company reported in a press release. "The vases, often filled with fresh, fragrant flowers, were used as air fresheners to help cover engine odors and the scent of passengers themselves in cars without air-conditioned interiors. The vases themselves quickly became decorative as well and were widely available in catalogs and hardware stores."

The flower vases soon became a selling point. The New Beetle, which sold eighty thousand models in the United States in 1999, was advertised with slogans such as "a work of art with side air bags and a bud vase." According to the company, "The inherent cuteness and quirkiness of the interior of the Bug and exterior design particularly appealed to women drivers and sales skewed 60 percent female."[9]

Appealing to women, however, isn't always something car companies want to do. Automobile companies fear what marketers and anthropologists call *gender contamination*, the belief that if a car or other product becomes too heavily associated with one gender—in this case women—it will hurt sales. Indeed, there is an old adage in Detroit that a woman will buy a man's car, but in no instance will a man buy a woman's car. So, when sales of a vehicle skew too heavily female, car companies get busy.

"Most automakers don't believe being successful with women is good," automotive historian and author Katherine Parkin says. After recognizing their early success with women, the company rejected female consumers and said to themselves, "Okay, we've sold it to all these women; now let's change it into a boy!"[10]

In the case of the Beetle, to try to reclaim a stronger market share among men, the company settled on a more "masculine" design with a "toughened-up" color design, and they ditched the vases. In ads rolled out nationwide, they proclaimed, "It's A Boy!"

But it is worth stopping to ask: How did cars come to be gendered in the first place? Why were there ever ladies' cars and men's cars? There are not, for instance, ladies' washing machines and men's washing machines. Or girl and boy iPhones.

A quick dip into the messy, sometimes incoherent, field of gender theory will provide some clues. One of the world's most famous feminists, Simone de Beauvoir, proclaimed that one is not born a woman, but rather becomes one, "On ne naît pas femme: on le deviant." Sometimes translated as "women are made, not born," the phrase set off a whole new school of feminist thought and decades of debate on the very definition of what it means to be a woman.

Any young girl who remembers watching her mother "put on her face" can attest to the magic. My mother transformed from ugly caterpillar to butterfly to fully fledged femme fatale with a puff of powder and a spritz of perfume. As a young girl, I spent each Saturday morning at the beauty parlor with her, and each week she left a different person than she came in. My mother was a broke secretary, but you couldn't tell that from the way she looked. She knew how to use clothes, hair, and makeup to project an image.

Taking to the extreme what Simone de Beauvoir said, and my mother somehow instinctually understood, Judith Butler and other theorists have argued that gender is not necessarily innate, but rather it is "constructed" and "performed" within known societal influences and constraints.

That is to say that gender has less to do with actual biologically determined sexuality and behavior and more to do with cultural, societal, and personal choices. But exactly how does this work when it comes to cars? And why would it matter?

First, it is important to note the swiftness with which cars came to dominate the American landscape. In the 1920s Americans owned more than twice as many horses as cars, but by 1930 that trend had flipped and there was now an average of one car for each family. [11]

"Within a single generation our rural and pastoral landscape was transformed into the world's most productive industrial machine," Leo Marx writes in his classic tome on the relationship between machines and nature,

The Machine in the Garden: Technology and the Pastoral Ideal in America. "The machine," Marx writes, was a "sudden shocking intruder upon a fantasy of idyllic satisfaction. It is invariably associated with crude, masculine aggressiveness in contrast with the tender, feminine and submissive attitudes traditionally attached to the landscape."[12]

Indeed, despite the swiftness with which Americans embraced the car, early motorists were often derided as interlopers. They were even assaulted with rocks and rotten tomatoes and taunted with the derisive comments of the day, such as "Get a Horse."

But in the battle for the streets, cars would soon overcome pedestrians as well as the horse, whose waste was often piled so high that women had to raise their skirts when they walked down the street. Originally engineered and created, at least in part, so that city folks could enjoy the countryside, cars and the highways they required would soon infringe in almost every way on the nature they were meant to deliver us to. "Machines," wrote Samuel Butler in his novel *Erewhon*, "serve so that they may rule."

The car would soon flatten both its critics and the American landscape with its unabashed hunger for miles and miles of highway. In so doing, it would change the very essence of what it meant to be human. As Georgine Clarsen points out in "The 'Dainty Female Toe' and the 'Brawny Male Arm': Conceptions of Bodies and Power in Automobile Technology," the motorized vehicle allowed women access to power as never before with dramatic effects on what it meant to be both male and female.

In the wake of the automobile, "gender, came to be seen not as natural traits, proceeding causally, masculinity from maleness, femininity from femaleness," writes Beth Bailey in *From Front Porch to Back Seat: Courtship in Twentieth-Century America*, "but as identities that must be acquitted, earned and constantly demonstrated."[13]

And what better way to demonstrate your masculinity than by successfully cranking up an engine, pushing a car out of a ditch, or roaring down the road at great speed? In a remarkably short amount of time, the car became the driver of modern masculinity. Modern women would soon be taken along for the ride.

Becoming a woman was no easy task for me. My mother died when I was ten years old, and with a newly married sister and a misogynistic brother and father, I had to figure things out on my own. Throughout high school, I was a good student and a talented debater, and upon graduation, I was admitted to the University of Chicago.

And so I headed south to college. Edith Wharton waxed poetic about how the car changed the entry into the city. No longer was it necessary to enter every city from down on the railroad tracks, your vista polluted by the shanties that invariably lined them. Now it was the majestic highway that laid bare the great city.

This was especially true in Chicago. After high school I drove that little punch buggy down the wide-open highway to what at the time was a great city but not a behemoth, a city Carl Sandburg called a "lusty young giant," a city that was full of promise for a young girl who knew literally nothing.

I was unprepared in almost every way to enter the great university. Despite receiving a good education at a public high school in Waukegan, I was no match for those from better-funded public high schools, magnet schools like Stuyvesant, or the elite boarding schools some of my fellow students attended.

The weather didn't help. My first year, 1977–1978, we had eighty-nine inches of snow. Snow, clean streets, and safety dominated the political discourse of the day and the politics. If you didn't get the town cleaned up, you were not likely to be voted in again in Chicago. Mayor Richard J. Daley, with his well-oiled political machine of ward bosses and union jockeys, was particularly good at this. The mayor when I was in Chicago, Michael A Bilandic, was not.

One particularly cold and miserable day, I started to skid at a stop sign, and the punch buggy accelerated into the middle of a four-way intersection. I was T-boned by a pickup driven by a nice man from Iowa. There was so much snow it was as if it had happened in slow motion. With the unmistakable sound of metal against metal, the little car was creamed. I wasn't hurt, but I was certainly reminded of the life-changing possibility of the automobile. My Beetle was done for.

I would stay only one year at the University of Chicago. Overwhelmed by the literary professors with heavy German accents and after a near-deadly brush with a class in astrophysics, I left. I put three boxes on the Lake Shore Limited and headed to New York University to study film and television. And for the first time in my life, I did not have a car.

SEVEN

New York City

Everyone who values cities is disturbed by automobiles.
—Jane Jacobs, *The Death and Life of Great American Cities*

New York City, despite being a town in which only a small portion of the population drives, is dominated by cars. I remember looking down from my New York University dorm room at 5th Avenue and 10th Street watching the traffic move through the grid—sometimes seamlessly, but more often than not, erupting into a cacophony of horns and curses as an overly aggressive driver blocked the intersection, creating a traffic jam that stretched for blocks.

In *Traffic: Why We Drive the Way We Do (and What It Says about Us)*, Tom Vanderbilt calls the intersection a "perfect arena for clashing human desire." Traffic engineers, he reports, have isolated over fifty points of potential conflict in every four-way intersection. [1] New Yorkers, I was certain, had found every one of them.

Fresh from the Midwest, I was both frightened by and in awe of the Big Apple. Living on the South Side of Chicago while I attended university did not prepare me for the relentless honking, virulent driving, or belligerent taxi drivers that together made New York such a scene. I was overwhelmed by the sheer density of the traffic, the flood of pedestrians, and the nause-ating smell of the diesel buses.

Traffic could be especially difficult if there was a visiting dignitary, or if the United Nations was in session. That's when the small streets in Green-wich Village swelled to try to relieve the activity uptown. Traffic is a lot like water that way; it seeps out in every direction until it finds its own level.

While at NYU I studied film and television, but I also studied photo history. From the old street photographers, I knew that the streets of New York didn't always look the way they did from my dorm room in 1979. Less than one hundred years before, the city streets belonged to everyone. I had seen the evidence of it. Children played there, vendors sold their goods there, and people rode their horses there. The roads were muddy and dirty—filled with horse manure and sometimes raw sewage when it rained—but they were available to all. "Streets are not a city's veins but its neurology," Adam Gopnik wrote in the *New Yorker*, "its accumulated intelligence."[2]

But in New York, as in other major metropolitan areas, the car changed everything. In the early 1900s, as cars became more and more prevalent, the roads were paved and the clashes between cars and pedestrians became heated and bloody. "Today the automobile represents freedom," Peter D. Norton writes in *Fighting Traffic: The Dawn of the Motor Age in the American City*. "But to many city people in the 1920s, the car and its driver were tyrants that deprived others of their freedom."[3]

One study estimates that over 210,000 Americans were killed in traffic accidents between 1920 and 1929, three or four times the death toll of the previous decade.[4] Women and children were especially vulnerable.

❉

"The motor age marked its arrival in the American city with asphalt parking lots and concrete highways," Peter D. Norton writes in "Street Rivals:

Jaywalking and the Invention of the Motor Age Street." "But before the city street could be physically reconstructed to accommodate motor vehicles, it had to be socially reconstructed as a motor thoroughfare."[5]

Language played an important role in the automobile's evolution. The verb *to park*, which once meant "to plant a tree or spread a patch of flowers," was repurposed to mean "a space for a car."[6] The car itself—once a plaything of the rich, which had often been referred to as a "pleasure car" meant to be taken on "joyrides," had to be redefined as a machine with value for everyday people, as a vehicle of progress destined to create economic gain and a better life for all. In short, the car and the infrastructure it required had to be presented as inevitable. Those who opposed the car had to be seen as technophobes, unable to cope with change and fearful of progress.

Beyond the selling of the automobile itself, the very idea of the automobile had to be sold to the American public, an effort spearheaded by the auto industry and their allies in the automobile clubs, whose members were fanatical about the invention. Automakers and their enthusiastic fans were the bullying tech bros of their time, and the relentless pursuit of technological progress came with a similar set of themes to those that surrounded the coming of the digital age. It was cool to like technology. Those who saw the downsides of technology and wanted to address them were, of course, much less than cool—even downright stodgy.

To better understand this dynamic and how it played out in the realm of the automobile, consider the storyline of the Pulitzer Prize–winning novel from 1918, *The Magnificent Ambersons* by Booth Tarkington. The young protagonist, George, is the scion of a decaying fortune who is steadfastly anti-car, so much so that he shouts that slur of the day, "Get a Horse!" to a driver stuck in snow.

But his love for the daughter of a man who is a manufacturer of the newly invented automobile is unrequited until, get this, the somewhat nasty young George gets his "comeuppance" by being run over by a car while trying to cross the street. Only then does poor George see the headlights and realize that cars are the future; he gets the girl and the fortune her father has made from manufacturing cars. In the film version of the novel, directed by Orson Welles and released in 1942, a newspaper headline screams of increased pedestrian deaths from automobiles, but those concerns go unheeded and in the end the crash that sends poor George to the hospital seems inevitable, even valuable: a necessary coming to his senses for poor old George. "Having set up this dream town of the good old days," Orson Welles said of the film, "the whole point was to show the automobile wrecking it."[7]

Some critics argue that the scene that does George in is reminiscent of a scene in Charles Dickens's *A Tale of Two Cities*, when the carriage of the Marquis hits and kills a young child. The Marquis tosses the father of the dead child a coin and blames the child and other poor people for being "ever in the way."[8] Dangerous and deadly driving, after all, has always been the prerogative of the rich.

Later traffic accidents would be blamed on myriad causes: faulty equipment, poor roadways, or a lack of driver education. The inevitable danger of the automobile would be explained away in technical and passive language that led us all to ignore the inherent dangers of the technology. In other words, mistakes were made. All of it added up to a fair amount of disassociation from the damage the automobile could cause and a lot of blame-shifting. Not only did the car change our landscape irreparably—running highways through villages and forcing pedestrians to the curb—the car introduced a good deal of abstraction into the everyday lives of Americans.

Where once the countryside rolled by from the horse-drawn carriage or the steam powered locomotive, the arrival of the automobile created that unique feeling of seeing the countryside whiz by at forty miles an hour. Engineers and scientists now speculate that it is possible to view pedestrians as fellow humans if the pace is slow, around twenty miles per hour. Above that, anyone outside the vehicle becomes a blurry, hard-to-distinguish object. Too abstract to personalize.

As George so painfully learned, the pedestrian is no longer just a neighbor walking quietly to work; the pedestrian is simply another obstacle—ever in the way, always underfoot. As a result, drivers sought to remove their fellow citizens from the roadways in order to dominate both the roadway and pedestrians. This occurred not just through legal maneuvering, but also through cultural bullying. "A 'jay' was a hayseed, out of place in the city; a jaywalker was someone who did not know how to walk in a city."[9] According to one early definition, jaywalkers were "men so accustomed to cutting across fields and village lots that they zigzag across city streets, scorning to keep to the crossing, ignoring their own safety and impending traffic."[10] *Jay* was therefore a derogatory term that came to mean something like "dirty hillbilly" in what some have called an early effort to shift blame for deadly collisions from drivers to pedestrians.

To curb the practice of jaywalking, in many cities police officers or Boy Scouts distributed leaflets to pedestrians who crossed the street in unruly ways.[11] Whistling policemen bustled offending pedestrians out of the street, shaming them as they went. Women were often easily embarrassed by such techniques and were consequently shouted or bullied out of the way and off the streets. One woman confessed that the ridicule that included a shrill blast of a whistle from a crossing guard "pierced her whole system" and left her cowering.[12] Another woman reported being bodily carried from the street and deposited back onto the curb.[13] Some women reported being arrested. Women, who already had a slew of reasons to feel uncomfortable on city streets, now had another one—they were literally being driven from the streets.

Remember the famous scene in *The Great Gatsby* discussed earlier when poor working-class Myrtle, the mistress of Daisy's husband, is hit in the street by Daisy, who is driving the mammoth Silver Ghost toward "death and the cooling twilight"? The crash leaves Myrtle's left breast "swinging loose like a flap," her body so wrecked there is no need to check for a heartbeat. The carnage and the message are clear: the streets are no place for a woman. As Daisy's friend Jordan Baker says, "It takes two to make an accident." Meaning, of course, that impoverished, stupid, cheating women should get out of the way of speeding cars or suffer the consequences.

No wonder traffic engineers now refer to pedestrians as "vulnerable road users." Worldwide, about 1.2 million pedestrians die each year in traffic accidents.[14] More men than women die in pedestrian accidents, but scholars are unclear why that is. It might be because men are simply out on the street more, or it could be that men participate in more dangerous activities such as drinking, or it could be that men are simply more aggressive while trying to cross the street. As the scientists like to say: more research is necessary.

What is clear and unmistakably true is that we have as a society embraced "car supremacy." Gregory H. Shill, associate professor at the University of Iowa College of Law, has studied how we have systematically enshrined, embraced, and prioritized cars over pedestrians, public transit, and overall public health. He sums it up this way in an article for the *Atlantic*, "Car Crashes Aren't Always Unavoidable."

> A century ago, captains of industry and their allies in government launched a social experiment in urban America: the abandonment of mass transit in favor of a new personal technology, the private automobile. Decades of investment in this shift have created a car-centric landscape with Dickensian consequences. In the United States, motor vehicles are now the leading killer of children and the top producer of greenhouse gases. Each year, they rack up trillions of dollars in direct and indirect costs and claim nearly 100,000 American lives via crashes and pollution, with the most vulnerable paying a disproportionate price.[15]

❉

The noted science fiction writer Ray Bradbury is from my hometown of Waukegan, Illinois. In his well-received short story "The Pedestrian," Bradbury fictionalizes something that happened to him in the mid-1940s. After dinner with a fellow screenwriter in Hollywood, the two decided to take a stroll down Wilshire Boulevard. A policeman stopped the pair and asked them exactly what they were doing. Walking was such a foreign act in a city that was focused on cars that a policeman found the two well-dressed men suspect. In his story, Bradbury re-creates a similar scene, except that the pedestrian gets arrested by an autonomous unmanned police vehicle. His crime: walking.

The car, after all, came to define everything around it. It created the pedestrian. One is defined in relation to the other.

When I was a child, wandering the streets of the hometown I shared with Bradbury, I couldn't leave the house without the admonishment to look both ways before I crossed the street. Perhaps it was the earlier tragic death of my father's brother that lingered over my childhood, or maybe it was the sound of the cars whizzing by in front of our house. No matter. I was fearless behind the wheel, but fairly timid out on the streets alone.

By the time I arrived in New York City in 1979, jaywalking was something of a sport for hardy New Yorkers who tested their own resolve and timing, cutting through traffic with an alert gait and a devil-may-care attitude.

Too timid to do my own jaywalking, I patiently waited for traffic to stop and then scampered across city streets before the lights turned, all the while pounding on the little buttons attached to light poles, which some refer to as "beg buttons."

Many years later what I and others long suspected of being true was confirmed; most of the crosswalk buttons in New York are not operational. Recently released data from the New York City Department of Transportation confirms that only 9 percent of all crosswalk buttons in New York are operational. The remaining 91 percent are set to fixed timers. As *Salon* magazine so aptly put it, these inactive buttons "simply serve as placebos for Type A personalities or germ-laden playthings for restive children." [16]

"Electronic signalization was created to allow for the fast flow of vehicular traffic," Shill writes, "and to prioritize traffic on main streets over that on side streets, and over pedestrians in general. Indeed, crosswalk signals are timed not for pedestrian safety or comfort, but for maximum vehicular throughput." [17]

While ants, bees, and birds have had eons to learn how to move across great distances while weaving in and out in an orderly, productive fashion, humans have had only about one hundred years to learn how to coexist with the automobile. In addition, the car can be best understood not just as a single, unique invention, but rather as the beginning of a new technological ecosystem, complete with parking, garages, stoplights, highways, and gas stations. Not to mention tons of unwanted effluent in the form of exhaust, gases, and leftover used oils and engine parts. To accommodate these technological changes and all the good and the bad that they created, humans had to learn new rules, new behaviors, and new skills.

"The machine's entry into the garden," Leo Marx wrote in *The Machine in The Garden: Technology and the Pastoral Ideal in America*, "presents a problem that ultimately belongs not to art but to politics." [18] Making way for the automobile meant reenvisioning city streets and building cross-continental highways. Often roads carved out for progress meant lanes that segregated White and Black neighborhoods—roads that, in the words of one activist, sent White roads through the Black man's bedroom. Everything about the car's entry into cities and towns was political—there was little art involved. These new highways created urban slums, trapping residents into decaying neighborhoods with little access to stores and creating what we now call food deserts for low-income residents.

Morever, some argue that African Americans and other minorities are more often ticketed for offenses such as jaywalking. In a well-known case, for example, in Orange County in 2020, a homeless Black man was killed by police officers after being charged with jaywalking.[19] California governor Gavin Newsom responded by signing the Freedom to Walk Act, making jaywalking legal in California.

❋

Cars were not a problem only when they were on the street, however. When not in use it was necessary to make arrangements for storing cars, which—except for cars used for commercial purposes—is most of the time. To quote the Nobel Prize–winning Czechoslovakian poet Czeslaw Milosz, "Even asleep, we partake in the becoming of the world."[20] When they are parked, cars are almost as big a problem as when they are moving, defining everything around them.

As a result, parking has historically been an issue not only for car owners, who had to expand their residences to make room for the automobile, but also for cities, which had to make room for cars both while moving and, then again, when they were parked. Indeed, the need for free or discounted, accessible parking was a major factor shaping urban and suburban design beginning in the early twentieth century and continuing today.

Remarkably, most street parking is either free or heavily subsidized. Free parking, argues Donald Shoup, in his fascinating encyclopedic book on the economics of parking, *The High Cost of Free Parking*, is essentially "charity for cars."[21] Shoup argues that "free curb parking in a congested city gives a small, temporary benefit to a few drivers who happen to be lucky on a particular day, but it creates large social costs for everyone else."[22] Shoup is such an icon there's a Facebook group devoted to the man and his ideas. They call themselves Shoupistas.

According to Shoup, free parking drives up the price of everything, as the price of the land and asphalt is simply baked into the cost of the products in the store. As the car gained traction, the need for designated parking became attached to each new building, thereby driving up costs even as it changed the landscape irrevocably. Moreover, the lingering cultural myth that women were poor drivers and even worse at parking led some mall developers to make parking spaces larger than necessary. "To attract female consumers and workers, the shopping malls needed women to drive, and their plans called for big spots for easy parking," Katherine J. Parkin states in *Women at the Wheel: A Century of Buying, Driving, and Fixing Cars.*[23]

So, for example, when a mall opened just outside of my hometown, the parking was robust, and many stores and shoppers relocated to what became a new destination—willingly, but unknowingly, absorbing higher prices as part of the deal. As was the case in so many towns, this left the commercial area to deteriorate. In Waukegan, the diaspora left the downtown free for us to drag race and "scoop the loop," but ultimately damaged the town's long-term development, as the ideal location for shopping and dining moved west toward the new mall and away from the traditional downtown center near Lake Michigan.

In New York, of course, there is free curbside parking in some neighborhoods. But consider how much time and gas are wasted while cruising for a parking spot. In one study, Shoup cites the average time to find a curb space on fifteen blocks on the Upper West Side of Manhattan as 3.1 minutes and the average cruising distance as .37 miles. These findings were used to estimate that the search for underpriced parking on these fifteen blocks alone creates about 366,000 excess vehicle miles of travel and produces 325 tons of CO_2 per year.

"Although cruising may seem to be an inevitable consequence of living in a crowded city, some drivers believe they have good 'parking karma,'" Shoup writes, "or an uncanny ability to find a curb space when they arrive at their destination. Give the laws of probability, some drivers will be luckier than others in finding a parking spot, and they may interpret this luck as a rare gift rather than pure chance."[24]

In the late 1980s I met a man who had excellent parking karma, although he claimed he paid off the parking fairy. He would later become my husband. His car was an old Toyota Corolla with a cassette tape deck. We played Tina Turner and Bruce Springsteen and spent endless hours on the Long Island Expressway traveling to a shared beach rental in the Hamptons.

The most popular novel at that time, Tom Wolfe's *Bonfire of the Vanities*, included its own terrible crash, quite reminiscent of those in *A Tale of Two Cities* and *The Great Gatsby*. The main character, a Wall Streeter, hits a Black boy with his Mercedes while driving his mistress home from the airport. All hell breaks loose, and the car crash and resulting trial light up the recurring themes of race, privilege, and wealth—as expressed through the automobile.

It's an old story. In the United States, about as many people are killed every year in car accidents as live in Houston, Texas. There is some evidence that minorities are disproportionately affected, as are the elderly, but everyone is vulnerable whether they are driving or walking the streets. Indeed, the very word *accident* has come to make it seem as though car crashes are inevitable and unavoidable, like thunderstorms.

The fact that accidents are common means only one thing: we've become accustomed to them. We've accepted the carnage as the price of our vehicles. We've privileged our own needs over the lives of many, many Americans. Indeed, as Shill, who has studied the ways in which car superiority became encoded into law, says, "We never even stopped to ask whether a different bargain was possible."

❋

On a recent trip to my hometown, I visited the intersection where my father's brother was killed at age six. After that, I went to the local library where, through the amazing feats of digitization, I was finally able to read a three-paragraph account of his death, which had occurred almost one hundred years earlier, in 1931.

> Donald Nichols, 6, was fatally injured Monday when he dashed into the street and was struck by a school bus just in front of the Oak Terrace School. He died of a basal skull fracture three hours later at Highland Park Hospital.
>
> Donald was standing on Prairie Avenue curbing just west of Pleasant Street, talking with young friends. Suddenly, they said, he dashed out into the street and was struck by the fender of the bus. As he fell to the ground his head struck the pavement.
>
> The child's death was the result of an unavoidable accident, a jury under Dr. M. J. Fenney decided Tuesday morning following the inquest at the Kelley funeral home in Highland Park. Police Chief Emil Lauridson of the city is considering the creation of a school patrol to prevent similar accidents in the future. [25]

So, it would appear that my father may not have been directly responsible for Donny's death. Was my father there and taunting his little brother? Did he push his little brother into the street, either accidentally or on purpose? Were they playing a game of pony and my father, who held the rope, pulled hard, causing his little brother to fall and hit his head before he was hit by the bus? Who knows? We will never know. And by the time I was old enough to hear my father's stories, he didn't really know either. It was just a quiet stream of lies and half-truths.

Out of all of that what we can surmise is this: Donny died doing what children have always done. He was an impetuous kid who asserted his right to the street during a period when the street was becoming contested territory. He was doing what nearly every boy of the time did, but the times were changing.

As I walked to the car from the library, I wondered: Did my father lay himself upon the cross and bear that guilt for all those years when it was just his misfortune to be caught in the crossfire of technological change? Did it start his unraveling? Could things have been different? It is said that my grandmother never recovered from the death of her small boy. And I believe that. She was a shell of a woman by the time I met her.

I remember once, when my own son was about Donny's age, we were out walking in the rain. We were just walking down the street the way people do every day. My son saw a puddle and he decided to jump it. He jumped straight into the path of an oncoming BMW. My heart stopped; the car swerved. My son cleared the puddle and then looked at me, terrified, suddenly realizing what he had done. I thought of my grandmother and my father and all the pain they endured. I hugged my son close and took him home.

Later he told me he was trying to jump the way a gazelle leaps. So, we had a new rule: no more gazelle leaps.

EIGHT

My Honda Odyssey: Minivans and Moms

That's what raging against the minivan has come to mean to me. It's the quiet rebellion against obsessing over the optics and outcomes of motherhood from the kind of car we drive to looking like we have it all together . . . Spoiler alert . . . as much as I raged against it, I lost my battle against the minivan. It now sits in my driveway covered in crushed Cheetos, smelly soccer gear, and remnants of my self-esteem.

—Kristen Howerton, *Rage against the Minivan*

For the most part, the stories that surround the creation of the American automobile industry revolve around men, speed, and American manufacturing might. But the origin story of Japan's famous Honda Corporation lays bare a different truth, one that reveals the car as a tool of domestic servitude and its role in shaping gender relations.

Honda was founded in 1947 by Soichiro Honda, an erratic and sometimes cantankerous entrepreneur and inventor, as a way to avoid helping his wife with her chores in postwar Japan. In the immediate aftermath of

World War II, Japan faced desperate food shortages. Rice was often in short supply and only available on the black market. During this period, Honda's wife, Sachi, became exasperated by her daily search for food through neighboring villages on an old bicycle. Complaining to her husband, who had just sold one of his previous companies and had taken on a bustling side business of making Japanese moonshine for his friends, she demanded that he put down the booze and take on the shopping himself.

"No way did I want to do that," Soichiro told a reporter and corporate biographer. Instead, he retrofitted an old bike with a motor made from parts he found on the war-ravaged streets of Tokyo and gave it to his wife to do her shopping. It was the first of many motorbikes and motorcycles the company would make, and it was the beginning of his manufacturing dynasty.

❋

Some fifty years after Soichiro Honda created that first small motored bike for his wife, I bought my Honda Odyssey minivan in Boston. That year, 1995, the company ran an ad for its popular minivan, an ad aimed squarely at young couples like my husband and me. Echoing the famous call that starts the Indianapolis 500, it said: "Ladies and Gentlemen, Start Your Families."

That's what we did. I was thirty-five years old when my son was born, in February 1995, and we picked up the van a few months after his birth. Our van was purple, but the salesperson called it eggplant. At the time, it was the best-selling minivan in America.

"Whatever the Honda Odyssey is selling," wrote *Wall Street Journal* car critic Dan Neil, "Americans are buying."[1]

Minivans are essentially repurposed cargo vans fitted with bench seats and an extra third rear seat for added passengers. My big purple van, however, was a far cry from the vans of my youth. Those most often had shag carpets and mood lighting and were outfitted with a mattress in the back. They existed primarily as a kind of mobile bedroom, complete with wine racks and bumper stickers that warned "If it's a rockin', don't come a knockin'."

The minivan I purchased was far more pragmatic and a lot less sexy. It was engineered so that women could more easily navigate their dual roles of working and mothering. The van, for example, helped a woman slide groceries into the back without having to bend too far over in the suit and high heels that she had worn to the office. Moreover, it included a single sliding door on the side to help load and unload children. If the car industry of my childhood had once been about speed and power, the minivan of my adulthood sat squarely on the side of domesticity.

The first minivan was introduced by Chrysler in 1983 in a bet the company move made by Chrysler CEO Lee Iacocca. Iacocca had been dismissed by Ford in 1978; he joined Chrysler that same year to try to keep the company solvent. At the time, Chrysler was teetering on the brink of bankruptcy after a series of missteps that included making and marketing large, gas-guzzling models even as consumers began seeking fuel economy. In addition, Chrysler, like other Detroit companies at the time, had heavily underestimated the power and appeal of Japanese brands like Honda and Toyota.

Iacocca was a charismatic salesman with serious "car guy" credentials. He was, after all, the man who had given us the Mustang my sister drove in her late teens. The minivan he introduced would single-handedly save the Chrysler Corporation from bankruptcy. In its first full model year, 1984, Chrysler sold more than 200,000 minivans. By 1988, this number had climbed to 450,000, and it peaked at about 500,000 vehicles a year in the 1990s.

Beyond being an exemplary business turnaround story, the minivan is also an emblem of changing American life. The 1980s was a decade characterized by vast changes in the social fabric. Women were starting to enjoy the greater freedoms, rights, and awareness fought for and gained during of the women's movement of the 1970s. And as the job market opened up for women, their paychecks bought them new clout as consumers. Women were beginning to believe that they really could "have it all" in terms of being both successful career women and homemakers. The minivan symbolized that change so indelibly that one is in the permanent collection of the Smithsonian Museum. "The minivan is a big piece of the American story in the '80s and '90s," a curator from the Smithsonian told the *Washington Post*.[2]

The minivan is a testament to a particular story about the way women worked and lived then and now. It also represents the fraught relationship that automobile manufacturers have always had with female drivers. Car companies needed women as consumers, but they needed to appeal to them in ways that were not overly threatening to male customers, who overwhelmingly identified with their vehicle's make and model.

Women had been identified as prime customers as early as 1920 by General Motors, and a detailed 1925 sales manual from Ford cited the influence of women on early car purchases. Later, car companies would pitch a second car as a way to get women to do errands men would rather avoid. A 1967 advertisement, for example, suggested that buying "the little lady" her own car would allow her to relieve men of their own chores. As the ad put it, "There's a lot she could be doing during the week if she had a car of her own. You know, the stuff you end up doing on Saturday."

This sentiment was, and continues to be, a familiar refrain in marketing automobiles. Women drivers made men nervous. They threatened male dominance at home, in the workplace, and in automobile culture in general. As a result, women's growing dominance as both drivers and consumers put automakers in a tough spot: they needed to market a car to women without alienating male drivers. They did this by manufacturing and promoting a "woman's" car that tied women more directly to home and hearth even as it roved about the suburbs.

"Sturdy, spacious, and utilitarian, the 'women's car' has long been recognized as the perfect vehicle for carrying kids and cargo," according to auto historian Chris Lezotte. "Thus, the ubiquitous station wagon of the 1950s and 1960s, the 1970 hatchback sedan, and, finally, the popular minivan introduced in the 1980s."[3]

As Soichiro Honda did decades earlier in Japan, car companies in the United States framed independence for women as the ability to fulfill their domestic duties more perfectly. The minivan, therefore, was simply a wildly profitable extension of Honda's earlier strategy of reconfiguring transportation as a domestic technology that helped women in their roles as wives, mothers, consumers, and caretakers. Men may have driven for speed or excitement, but make no mistake, women were supposed to head straight to the grocery store. "Despite changes in form, style, and cargo space over the past 60 years," Lezotte concludes, "the function of the woman's car has remained the same. And that is to firmly reinforce women's gendered roles as wife and mother."[4]

By the time I bought my Honda Odyssey minivan in 1995, I was working as an editor at *Harvard Business Review* writing about women and work. Profiling, for example, the women who had gone into the vast automotive plants just as the men flooded out of them to fight in World War II. Rosie the Riveter emerged from the kitchen to build the machinery necessary to fight and win that war, all while looking stylish in a painting by Norman Rockwell that was also turned into a piece of war propaganda by J. Howard Miller, which famously proclaimed "We Can Do It!" During the five years from 1942 to 1947, when women like Rosie were on the shop floor at plants like Ford's River Rouge Complex, productivity rose, product cycle time dropped, and quality improved. Yet despite her success, Rosie was forced off the factory floor when the war ended, her achievements buried in the books and wiped out of our consciousness. She had proven her abilities, but she remained a cultural anomaly: a woman in a man's job.[5]

Work, especially factory work at that time, was overwhelmingly seen as belonging to men. As wartime production ended and auto production resumed, manufacturers claimed that while the Rosies on the factory line knew how to rivet, they did not know how to weld, and there was no hope of teaching them how to do so, despite the speed at which they had earlier learned their factory jobs. Rosie was sent home.

As the daughter of a single mother, the basic injustice of Rosie's story resonated with me. I knew that women often needed to work and that women like Rosie and my mother were supremely competent. Yet even as I was writing about the challenges women have faced in the workplace since World War II, I was coming up against my own set of struggles. I was openly paid less than male editors at the same level while trying to balance raising a small child.

Despite the well-argued pleas for equity in the workplace and at home that I published in the magazine, my life was more servitude than I cared to admit. I rushed out early in the morning in my big plum van to get my son to day care and then, at the end of the day, fled the office in a panic to pick him up and make dinner. After I got him to bed, I would return to my work, marking up long galley sheets of charts and graphs that would accompany detailed articles on strategy and finance. I was exhausted.

I was not alone in my plight.

Earlier technological revolutions that were supposed to make women's workload lighter had failed to produce the time savings that manufacturers had promised. The car would prove no exception.

In *More Work for Mother: The Ironies of Household Technology from the Open Hearth to the Microwave*, historian Ruth Schwartz Cowan explains the irony. The vacuum cleaner, the washing machine, the refrigerator, and the car, she suggests, changed the household system, but they did not save women time. Instead, they replaced labor that the average housewife might have employed to help her, including delivery boys, laundresses, and cooks. In short, even as women gained appliances and automobiles, they lost paid household labor and delivery services.

Moreover, as these appliances were introduced into the household, cleanliness standards increased dramatically, often driven by manufacturers and other commercial interests. After the dishwasher became routine, a drinking glass pulled from one had to be spotless, whereas a small smudge or slight cast had been the previous norm. Clothes were laundered after each wearing and whites had to be pristine, when in the past a dress or pair of dungarees might have been worn several days in a row. These higher standards put mothers on the hook for a whole new level of domestic performance. As it turned out, the technological revolution of the early twentieth century made women's work less visible, but not less time-consuming or demanding. As Cowan describes:

> By midcentury the time that housewives had once spent in preserving strawberries and stitching petticoats was being spent in driving to stores, shopping, and waiting in lines; and the energy that had once gone into bedside care of the sick was diverted into driving a feverish child to the doctor, or racing to the railroad station to pick up a relative, or taking the baseball team to the next town for a game. The automobile had become to the American housewife of the middle classes, what the cast-iron stove in the kitchen would have been to her counterpart of 1850—the vehicle through which she did much of her most significant work, and the work locale where she could most be found. [6]

Of course, the situation would only get worse as the twentieth century came to a close. By that time, driving had become a major part of women's work. On average, American women spend about an hour behind the wheel each day. Although men drive more miles, women are more likely to have passengers in the car when they drive. In other words, they are often taking someone somewhere or picking them up. "If you look at trip rates by male versus female, and look at that by size of family," Tom Vanderbilt writes in *Traffic: Why We Drive the Way We Do (and What It Says about Us)*, "the women's trip rates vary tremendously by size of family. Men's trip rates look as though they don't even have a family."[7]

That was the issue so many women of the Baby Boomer generation faced. They grabbed the golden ring of professional jobs and got sucked into a whirlwind of work, ferrying, and commuting.

About the same time that I bought my minivan, Martha Beck wrote her influential book, *Breaking Point: Why Women Fall Apart and How They Can Re-Create Their Lives*, in which she set out the fundamental Catch-22 for women. "A woman who spends all her days doing housework holds a lowly social status, but if she does anything else, she is neglecting her 'proper' responsibility and receives social censure."[8]

Certainly, my generation held the view that mothers were supposed to sacrifice themselves for their families. Baby Boomer women had a new dream of excelling personally and professionally and, in many cases, as I did, providing financially as well. This double bind, still prevalent today, has the power, Beck wrote, "to literally destroy women's lives."

If Betty Friedan, author of *The Feminine Mystique*, famously wrote about women in perfectly decorated homes bored out of their minds, Beck characterized them as being stressed beyond belief behind the wheel. In Friedan's day, women took tranquilizers, also known as "mother's little helpers." By the 1990s, women were given an overwhelming number of prescriptions for antidepressants and sleep-aid medications.

The problem is structural. There simply are not enough hours in the day for all that women must do. "Working mothers," Ann Crittenden writes in *The Price of Motherhood: Why the Most Important Job in the World Is Still the Least Valued*, "put in longer hours than almost anyone else in the economy. On average, they are estimated to work more than eighty hours a week.

Time-use surveys confirm that as women enter the workplace, they take on the equivalent of two full-time jobs, forcing them to cut back on everything in their lives *but* paid work and children."[9]

Automakers set out to capitalize on this time crunch for women by turning the minivan into an additional domestic space, complete with video screens, study trays for kids, cupholders for drinks, vacuums for spilt food, and, in one prototype, a washer and dryer that could be used at sports practices. Fast food restaurants and the makers of packaged food were quick to notice the trend and began testing burritos for one-handed convenience, so they could be eaten while driving, and making yogurt a sippable drink instead of something likely to slip off a spoon into the netherworld between the back seats of a van.

With so many obstacles to surmount, it should come as no surprise that Honda named their popular minivan model the Odyssey, after the famous Greek epic poem. *The Odyssey* is about the tortured journey of Odysseus, king of Ithaca, who is trying to return home after the Trojan War. He encounters many trials, including a goddess who takes him hostage for seven years; a visit to the Lotus Eaters, who give his crew a strange fruit that makes them forget their desire to return home; and raging storms that cause his ship to veer off course. It is an epic adventure tale that is, at heart, about the struggle to get home. The women who drive Minivans understand this kind of epic journey.

The minivan, after all, helps fuel the already hectic lives of women. "Mothers tend to form their lives and schedules around the needs of their children," the director of corporate research at Chrysler told the *Washington Post*. "Children need a vehicle they can be picked up in, change clothes in, and be taken to the next activity in, and the minivan is it."[10]

Marketing studies of the time revealed that *racing*, *juggling*, and *balancing* became the key words that women used to describe their lives, verbs that in the past were more likely to be used to describe circus acts than motherhood.

More recent minivan models have reinforced this trend by offering closed-circuit monitoring of the back seat, an in-car intercom, and apps that allow drivers to adjust the air conditioning settings in the back. It is a vehicle with so much functionality that one car critic called it the Swiss Army knife of the auto industry. As such, it has become a new private domestic space that, as one automaker claimed, is more affordable and advantageous than a remodeled basement.

"Minivans are marriage savers," *Wall Street Journal* writer Dan Neil once wrote. "The primary cause of intimacy issues among American couples is daily stress and aggravation, which scientists call children. Minivans reduce the strain on the harried, horned-up bill-payers in front, by making things quieter, smoother, easier, safer, more connected and less contentious in the back."[11]

While they do reduce tension in the car, they are often small prisons for the women who drive them. Perhaps no other vehicle has ever been so loved and hated at once. Glennon Doyle famously raged against hers in her bestselling memoir, *Love Warrior*:

> I arrive at my minivan. It's still sturdy boxy and gray—just as I left it. But as I put my hand on the door, I taste hot hate in my mouth again. I pull back and realize quickly that I despise this minivan. I take a step back to gain some ground between us and I stare while the van morphs into a symbol of my decade-long loyalty, sacrifice and naïveté. The van screams: *I am wife! I am a mom! This is who I am! I might not be flashy, but I love my life!* Everything about this minivan is proof that I'm a fool.[12]

The minivan proved to be such an important signifier of what it meant to be a woman and a mother that, by the year 2000, *Businessweek* reported that both Al Gore and George W. Bush were wooing a new demographic called "Minivan Moms." Minivan Moms were similar to soccer moms in that they were highly affluent, well-educated, and considered swing voters. Cars have been used for political purposes by women since their invention, most famously by suffragists, who staged rallies and literally shouted from the rooftops of their cars to win the vote. But Minivan Moms were the first group of voters identified by their car models.

I attended the Million Mom March in Washington, DC, in 2000 with my minivan. Like so many parents, I was outraged by the carnage that gun violence causes and gave up a precious Mother's Day to protest the power of the gun lobby on Washington.

At the march, I purchased a small bumper sticker that I thought triumphantly proved my point. "I am a mom and I vote," it said, next to a picture of a crossed-out gun. I pasted it on the rear bumper of my van. It was my very own roving political statement.

By affixing that bumper sticker on my big plum minivan, I joined a long list of women who have used their cars for political purposes. Suffragists from Illinois, for example, embarked on a series of road trips in the state during the early 1900s, using the car as a public platform and an object of what historians have called "ritual decoration."[13] Early cars did not have bumpers; instead, the suffragists attached handmade pieces of fabric to their vehicles, beginning a long-lasting trend of making political statements with automobiles.

While I was particularly proud of my small moment of political activism, my bumper sticker didn't please everyone. Later that year, my husband took our son on a camping trip to Mount Greylock in western Massachusetts. While they were away from the campsite, someone incensed by my pro–gun control sticker keyed my minivan with this mocking message: "Retard, vote for that."

I drove that car for another 100,000 miles with the word *retard* etched into the side under my driver's side window.

✳

"The Machine develops—but not on our lines," wrote the English novelist E. M. Forster, who was astonished by the automobile's power to shape human behaviors and society. "It proceeds," he wrote, "but not to our goals."[14]

Even as we make and choose our technology, technology is busy shaping us. Yet the life-changing nature of the car and all it represented wasn't always apparent as we were donning suits and heels and hopping in a van to get our work done.

Most women do not consciously choose to spend so much time raising their children in the car. Pregnant women certainly don't envision feeding or supervising their offspring at sixty miles an hour. As someone who has written about the history of women and work, and as someone who has lived it, I can attest that women never set out to create lifestyles that would drain them of sleep and spirit.

After all, I wasn't shopping for a minivan that fit my needs; I was buying something that worked well for my family. But just like Soichiro Honda's wife, who didn't want an electric motorbike, but *did* want help with the chores, a minivan was the best I could do. No one was going to help me with my chores or my work—not even the minivan.

❊

In 2003, both my minivan and my son were eight years old. My Mustang-driving sister had died of ovarian cancer just after Jacob was born in 1995. She saw my son just once. I hobbled through O'Hare post–cesarean section with a tiny baby in tow, determined that she would see him before she died.

As she was dying, she asked me to write about the toxic illnesses in our hometown of Waukegan, Illinois. Sue and I had grown up in a heavily polluted town, and her illness had made us both question the role it might have played in our health. There were three Superfund sites in my hometown and factories galore down by the lake where we used to swim. There were paint factories, asbestos product manufacturers, steel companies, and a coal-powered gas plant. We raced past those factories a thousand times in her Mustang, inhaling the smoke from the stacks that towered over the town.

Later, after my sister had died, I began to write about the toxic legacy left behind in Waukegan when the factories closed and the jobs went overseas. I once did a story about the man who ran the Manville plant in my hometown, which manufactured building materials. He told me every single member of his staff had died from workplace exposure to asbestos.

Ovarian cancer is a highly heritable disease, meaning it sometimes runs in families. So, after Sue died, I became a patient at the high-risk clinic at the Dana Farber Cancer Institute in Boston and began working in earnest on a book about the pollution in our town. I was about halfway through the book when I drove my old plum van down to the hospital for my yearly screening. My worst fears were confirmed when the radiologist ran his sonogram over my belly and gasped: "What the hell is that?"

I had a four-inch cancerous pancreatic cyst that needed to be surgically removed. The day before the surgery I ran a bunch of different errands and then spent the rest of the day vacuuming the big plum minivan, cleaning the windows, and wiping down the seats. I wanted to restore order, to make the vessel perfect again.

I once read a story about a woman who cleaned and buffed every part of her minivan before walking to her death in the open surf. Somehow, I understood that woman. I suppose I thought if I was going to die, I wanted to die with things in order. Maybe that's what she thought also.

Cleaning, after all, is the defensive action of all moms in trouble. After two surgeries and thirteen weeks of chemotherapy, it became clear that I could survive my disease. What would not survive was the minivan. Dozens of drives to the hospital and plenty of school pickups proved that the van did not have the kind of luxury suspension I needed. Each drive in the van post-surgery sent shock waves up my abdomen; each stiff turn led to new bouts of nausea. As much as I cherished that car, enjoyed every rattle, reminisced about every crayon mark, and smiled at every old Cheerio it coughed up from its seats, I needed to let it go. I needed to say goodbye. It was time to move on.

NINE

Cars, Consumption, and Cancer

There is only one efficient speed: faster; only one attractive destination: farther away; only one desirable size: bigger; only one rational quantitative goal: more.

—Lewis Mumford, *The Myth of the Machine*

In our role as consumers, we are not always as rational as we hope to be. We are not clear about bringing our values in line with our purchases, our fears about global warming in line with our choice of automobiles. I bought my sport utility vehicle (SUV) at the same time I worked on a book about the industrial pollution in my hometown and the diseases that I believed it had caused in my family, including my own bout with cancer. I was aware of the environmental damage cars and the infrastructure they require inflict on the environment and of the excessive engineering and use of materials that are part and parcel of all automobiles, but especially SUVs.

So why did I buy a Honda Pilot in 2005? My SUV had plush leather seats, and when I drove my son to school, I could hit all four of the speed bumps on the way without pain. I used it to drive unruly but lovable boys to school and soccer games and, no matter how many times they kicked the back of my seat, I did not flinch.

To be fair, while I had a good reason for wanting the Pilot, I was also somewhat in awe of the big beast. It was a lively blue, it had welcoming bucket seats, and it had storage everywhere. I could put a cup of coffee and a water bottle up front. Its whiff of luxury was a comfort since I was daily in line with a veritable lineup of luxury cars when I went to pick my son up from private school. It wasn't exactly showy, but it certainly didn't make me seem like the poorest mom in town.

I guess I have to say that the car called to me. And, yes reader, I bought it. Or, rather, my husband did.

Years of watching my dad in the showroom made me skittish at dealerships and slightly nauseous just thinking about negotiating the purchase of a new car. So, I did what all car-loving feminists do: I asked my husband to do the negotiating and sign for it. To this day, I do not own a car. All our cars are now and have always been owned by my husband, even if they are, in every other way, "my car."

❊

My 2005 Pilot was an extreme example of the ways that cars can lead us to make irrational choices both personally and as a society.

As early as the 1930s, important voices within the car community were arguing for restraint and common sense in car design. In 1938, automotive engineer Delmar G. Roos laid out a clear brief for the automobile when he wrote: "The object of an automobile is to transport a given number of people with reasonable comfort, with the least consumption of gasoline, oil and rubber, and for the slightest operating cost and prime price."[1]

Around that same time, a staffer at the Automobile Manufacturers Association noted rather famously what should have been obvious. That is, the foolishness of using "two-tons of automobile to transport a 105-pound blonde."[2]

My Pilot weighed in at just around four thousand pounds. By the end of thirteen weeks of chemotherapy, I weighed about 120 pounds and, although I was a brunette, the math had not changed much since the 1930s. *Foolish* was one word for it, but like so many before me, I fell for the myth of the automobile—in this case a two-ton SUV with a sunroof and a fold-down third row. On my first solo drive, I hit the curb at a rotary and took out a small azalea bush, an incident that became the only off-road adventure I would take in that SUV.

The remarkable thing about this purchase was that I knew better. In his day, my father had excelled at extolling the small virtues in each new model of the Dodge Dart. At one time he sold more Dodge Darts than any other man in the state of Illinois. I had seen my father sell blue-haired little church ladies on an entirely unnecessary but equally alluring new interior color scheme; I had watched men who could barely feed their families buy the new truck, despite the strain it could cause in their marriage and the very real possibility that the factory would close and all would be lost: the job, the truck, the house, and the marriage in one fell swoop of bad luck, rising interest rates, and the resulting, if predictable, economic turmoil.

Moreover, as an editor at the *Harvard Business Review*, I had worked with many of the best minds in business; I knew for a fact that the car industry single-handedly launched our never-satisfied consumerist lifestyle with its invention of the annual model, in which each year's version was only marginally better than the previous one, but magically more alluring.

"Most of the time, people are irrational, emotion-driven, susceptible to outside influence, and largely oblivious to what's going on," historian Tom McCarthy writes in "The Black Box in the Garden: Consumers and the Environment." "Even when we fathom something of what's going on, we find it difficult to change our behavior. That's great news for advertisers and campaign managers, but not necessarily for society or the environment—because it means that we are less the captains of our own ships than we imagine."[3]

I may have been a tiny brunette riding around in a big SUV, with a load of boys and a bag of dirty soccer balls, but I was far from captain of my own ship. Instead, I was just another mother participating in the great consumer lie that is car culture.

✱

The auto industry—in particular General Motors' master strategist, Alfred Sloan, first introduced the marketing strategy of planned obsolescence in the 1920s. As the car market became saturated in the 1950s and 1960s, both the company and the industry in general implemented this new strategy with enthusiasm. They did this by proliferating the number of new models and by making cars less reliable over the long term—a one-two punch that more or less required drivers to upgrade.

Cars were made "ever more shoddily," according to historian James Flink in *The Car Culture*. He studied the number of older models on the road year after year and kept watching as the average length that consumers kept a new car fell precipitously. He claims, and many others concur, that "the decrease occurs with a regularity that makes it look very much as if it were planned that way."[4] Flink notes that "cars in the 1960s were delivered to owners with an average 24 defects each," a fact that led to greater frustration and a need to swap out the car sooner rather than later.

The purposeful proliferation of new models combined with the hassle of an older car caused consumers to "trade-in and trade-up."[5] Moreover, consumers were encouraged to take on debt to upgrade. This was in sharp contrast to earlier negative views of consumer debt.

McCarthy explains it this way: "When a study commissioned and publicized by General Motors in the late 1920s showed that the rate of default on consumer automobile loans was much lower than people had imagined and when the Great Depression taught Americans the economic importance of consumer spending, much of the moralizing based on older values and beliefs stopped."[6] And that's how consumer debt for rapidly depreciating objects such as cars and home appliances became normalized.

"Installment selling had come to stay!"[7] thundered one enthusiast who had studied General Motors' practices.

What was good for General Motors, however, wasn't necessarily good for the American consumer. In a dizzying downward spiral, more debt meant more work, which meant more driving to work, which meant more wear and tear on the car and a greater likelihood of costly repairs.

Consider this statement from a woman in the 1940s extolling the virtues of her new car even as she rationalizes what it will cost her in time and money:

> We did it: the new car is in the drive. All the neighbors are peering out their windows and can see that we're preparing for a little weekend excursion. Yes, sir we've done something for ourselves, we want something from life. After all that's why we are both working, my husband in the plant and I as a secretary for my old company.[8]

As this quote suggests, women were both the drivers of car consumption and often the bearers of the cost in terms of longer hours at the job made necessary by car payments. Cars shifted our supercharged consumerist lifestyle into high gear. "The automobile was slowly invested with the emotional aura of consumption," according to Wolfgang Sachs in *For the Love of the Automobile: Looking Back into the History of Our Desires.*[9]

Women, for their part, when not prohibited from using credit, seem to have taken up the challenge of high payments in return for the promise of unfettered personal mobility. An ad that appeared in *New Century Teachers Monthly* in June 1924, for example, told the story of a stenographer, Georgia

Greene, from Murray, Utah, who after saving for the down payment on her $500 Chevy was left with monthly installments of $39 each.

This rather determined young woman hustled up four passengers to carpool to cover approximately $32 of her costs. Glossing over the fact that this still leaves her $7 short and doesn't cover the costs of fuel, repair, or wear and tear on her vehicle, the ad proclaimed: "Her Chevrolet will really cost her nothing." The ad then chided her male counterparts who apparently lacked her hustle: "Can she really be smarter than one million men?"[10]

To spur sales and keep the market for new cars robust, General Motors started its own financing arm in 1924,[11] and Henry Ford, who at first reasonably opposed selling cars on credit, but later capitulated, started Ford's financing arm several years later.[12]

In his sweeping review of credit and American consumerism, historian Leon Calder reports that by 1930 "virtually all retailers of durable goods had developed their own time payment credit systems."[13] Even at that early date, 60–75 percent of automobiles were financed on credit, often over periods of one to two years. Or, as the catchphrase went at my dad's dealership, no money down and twenty-four easy monthly payments.

Today both the amount borrowed for a new car and the length of the loan have become bloated beyond recognition. Recently, the average new vehicle loan hit an all-time high of $40,290, with payments topping out at around $780 a month on a sixty-nine-month loan, according to the credit monitoring company Experian.[14]

As a character in Mark Twain's first novel, *The Gilded Age*, says gustily: "Beautiful credit! The foundation of modern society."[15]

✳

Cars are always tokens of their times—two-ton status symbols. "In the city of Zenith, in the barbarous twentieth century," Sinclair Lewis wrote in his

1922 novel *Babbitt*, "a family's motor car indicated its social rank as precisely as the grades of the peerage determined the rank of an English family."[16]

By the 1950s, the car's place in a rigid social hierarchy had become even clearer. The auto historian John Rae has written that in the 1950s there were "communities in which eating with a knife would have been more acceptable than driving anything but a station wagon."[17] Reading that line it is hard not to conjure up the moneyed enclaves of Darien, Connecticut, or Lake Forest, Illinois.

Fast forward forty years to my suburban Boston community, where driving an SUV to soccer practice on Saturday morning wearing a fleece vest from J.Crew and holding a Starbucks cup was our status statement.

Beyond the desire to fit in, research shows that men bought an SUV for its macho image and for the dream that it held out, however fleeting, of a rugged outdoor adventure just around the bend. Women craved safety—even though the high balance point of early SUVs made them likely to roll—and an escape from the minivan, which hit hard at every woman's sex appeal. No woman really looked hot climbing out of a minivan, after all.

But the SUV had another advantage for small women like me, who perceived themselves to be dominated on the road and in life. It gave us a fighting chance in cutthroat traffic. Boston drivers like to brag about how dangerous and chaotic our traffic is. They routinely and proudly refer to themselves as Massholes.

The ability of any vehicle to do damage to another is a feature that traffic researchers call a vehicle's "aggressivity rating."[18] SUVs in particular have the ability to damage other cars, and that gives those who drive them an edge in traffic.

As women, we felt our fragility in those days, despite the power suits, the big jobs, and the cars that were way too big for us. We were generally pushed around and underpaid at work, overworked at home, and tasked with making sure our kids got on the right teams, went to the right camps and, ultimately, to the right colleges. All of this entailed a good deal of driving. According to *Car and Driver* magazine, women drive about ten thousand miles a year.[19] And for women who do most of the ferrying of children, that can often mean fighting for space on the road at all the worst hours of the day.

SUVs gave us some muscle. For once, I could look out over men in smaller cars and not have to yield the road to a more aggressive driver. "The archetype of a sport utility vehicle reflects the reptilian desire for survival," according to Keith Bradsher, author of *High and Mighty: The Dangerous Rise of the SUV*.[20] We convinced ourselves those cars were safer, even when we knew better and even as we drove them ever more aggressively.

Ironically, I spent much of my time in that car listening to the rosary on CD. Marrying a Jewish man meant going to church alone, and I didn't want to do that. Instead, I ordered a set of CDs of the rosary, and I played them over and over in my car. That car quite literally became my church.

"Cars today are almost the exact equivalent of the great Gothic cathedrals," wrote the French philosopher Roland Barthes in 1957. "The supreme creation of an era, conceived with passion by unknown artists, and consumed in image if not in usage by a whole population which appropriates them as a purely magical object."[21]

In the stunning category of you "just can't make this shit up," comes the fact that Barthes was hit by a laundry truck after stepping off the curb in Paris just after lunching with French president Françoise Mitterrand. Barthes died weeks later of his injuries. Writing of his death many years later, the *New Yorker* called it *un mort imbécile*.[22] Or, the stupid death of a very smart man.

Barthes would warn us against making inferences about his writing based on the circumstances of his life—he was, after all, a brilliant philosopher and writer—but it is hard to believe his collision with the laundry truck wouldn't have affected his feelings about the car.

But why did he liken a car to a cathedral in the first place?

Because the French worship a car called the Citroën DS. Reportedly, *DS* stands for a "different spirit" or "distinctive spirit," but can also mean "goddess" when pronounced quickly in French. *Goddess*, which invokes both the feminine and the spiritual, is the nickname that stuck.

The car made its debut at the Paris Motor Show in 1955, where a record eighty thousand Frenchmen placed orders for it in ten days.[23] A low-slung, sexy car, the Citroën is in the permanent collection of the Museum of Modern Art and is fondly viewed as the vehicle that pumped heart and soul back into the French after the Germans knocked the hell out of them in World War II. Barthes waxed poetic about the DS in one of his essays, comparing its seamless construction to Christ's robes. For Barthes, the DS was a move away from the sheer power of automobiles to a more feminine and artistic vehicle.

"It is possible that the Déesse marks a change in the mythology of cars," Barthes writes. "Until now, the ultimate in cars belonged rather to the bestiary of power; here it becomes at once more spiritual and more object-like."[24]

That was what my Honda Pilot was to me. It was in every way a monster of a vehicle, but, quite literally, it was also the only place and time I had to pray alone. And I did it every single day. It was my spiritual home even as it ventured to every soccer field in eastern Massachusetts.

Soon that sacred space would become critical.

❊

Nothing can prepare you for the worst, and no car can protect you from heartbreak, no matter how many times you say the rosary in it and despite what the ads say. I was devastated when my son Jacob was diagnosed with leukemia at age fifteen in 2010.

Just days before, I had driven him and some friends in my Pilot to a weekend soccer game. He had scored once in that game. He had fielded a ball with his left foot dropped onto his back and shot it into the corner of the opposing goal that lay behind him, a move called a bicycle kick. Watching his ballet-like movements, the crowd caught their breath in a collective gasp before breaking into applause.

On the day of his diagnosis, after a difficult call with his pediatrician, we piled hurriedly into the Pilot, and I took him to Children's Hospital, where we formulated a plan to fight his disease.

My son is in every way a fighter. Fierce emotionally, physically, and intellectually, he looked at me with his wide brown eyes and said: "Mom, now there will be two survivors in the family." And there are.

But that was not at all assured at the time, and I knew that all too well, even if he hadn't quite grasped it yet.

According to Sachs, "However acute in vision, one would be blind who stood before a cathedral and saw nothing more than a shelter from wind and weather for the faithful. And one would be just as blind who stood before a shimmering automobile with its engine humming and saw only a vehicle for transporting people and their goods."[25]

Every time I got in that car that year and for several years after it, I prayed the rosary—begging, pleading, and then thanking God for my son's life, all at the same time. My Pilot was in every way my cathedral. A very direct expression of all I hoped for, both for my son and for myself. But I was not blind to its environmental costs.

❈

In 2008, two years before my son's diagnosis, I had published the book my sister had asked me to write about the possible links between the pollution in my hometown and her death from ovarian cancer and my own pancreatic cancer. Waukegan, Illinois, contained three Superfund sites, one of which contained over one million pounds of polychlorinated biphenyls (PCBs), highly carcinogenic chemical compounds. As one scientist I interviewed said, "Every chemical we know to be harmful to human health is in one of those sites."

Trained as a journalist, I had studied the science, had specialists review my sister's slides, and spent long hours studying the effluents from the factories and the littoral drifts of Lake Michigan. Thoroughly researched, my book would win an honorable mention in the Rachel Carson Book Award Contest sponsored by the Society of Environmental Journalists. The US Secretary of Labor at the time, Hilda Solis, honored me as a leader in the field of environmental health. Being right never felt so wrong.

Sitting in the studios of National Public Radio, I winced as an expert in environmental health from the Centers for Disease Control praised my work. Far from being a crank, I had only shown in a painful and personal way what many scientists and doctors either know or suspect to be true in one way or another: hundreds of chemicals in use today are capable of causing cancer in humans. Moreover, some of these chemicals can cause changes in the genetic makeup of children in the womb, switching on and off genes, creating epigenetic changes that may carry on for generations.

Besides their ability to cause cancer, toxic chemicals are likely the culprits in a surge of learning disabilities and attention problems in children. Some of these chemicals can also be linked to infertility. Air pollutants are undoubtedly and inextricably linked to an increase in asthma in children.

Still, even as a so-called expert with a book on the subject, someone who spoke to large groups of women about the health risks of pollution, I somehow had contained my own fears and anxieties. I wanted my book to be a barricade against the power of the pollution in my hometown and in our everyday environment. I wanted those risks to be codified and sealed somewhere where they had no ability to harm anyone. As the writer Alice Munro put it: "I wanted death pinned down and isolated behind a wall of particular facts. Not loose and roaming around."[26]

Still, I have not and will not research the possible causes of my son's disease; it's too painful. His disease could have had its roots in my own exposure or my mother's. He also had his own multiple exposures through air, water, and food, and the chemicals spread on the many soccer fields on which he played. I know too that car exhaust contains a multitude of chemicals capable of causing cancer—one of particular concern is benzene, a chemical that has been studied for its possible links to childhood leukemia.

Cars have been major sources of pollution almost since their invention, but definitely since 1921, when lead was added to gasoline to prevent engine knock. The additive has been known to be both unnecessary and dangerous almost from the beginning. Safer options existed, including ethanol, but they were rejected by the auto industry.[27]

Indeed, a woman from Harvard Medical School named Alice Hamilton—the first female faculty member on the esteemed campus—aggressively campaigned against the additive that cumulatively has affected intelligence in hundreds of thousands of children and led to violence in many others. No matter. As a woman, she was not listened to.

Despite early and often widespread concerns, the car has fumed, spilled, and oxidized literally hundreds of chemicals into air, land, and water for roughly one hundred years. Consider for a moment the "new car" smell that filled my nostrils for all my youth and much of my teens. It was likely created by the oxidation of numerous sealants and adhesives and the off-gassing of plastic mats and vinyl seats. These are referred to as volatile organic compounds (VOCs) and can include dangerous chemicals such as benzene and formaldehyde.

"We think long and hard about [scent]," Eric Mayne, spokesperson for Stellantis, the global automaker comprising Fiat Chrysler Automobiles and France's PSA Group, told *Car and Driver* magazine in July 2021. "We have a panel of people who monitor new car smell."[28]

And yet the problem is not as simple as it sounds.

Remarkably, some people like the dangerous scent very much, as it reminds them of the prize of a new purchase; others not at all. In China, for example, consumers eschew the smell for fear of its health consequences.

J. D. Power, the company that ranks automobiles, reports that complaints about new car smell are ten times higher in China than in the United States.

Here in the United States a company called Chemical Guys makes an air freshener that mimics the scent and is used by detailers hoping to assure their customers they have done a good job. Or by those who simply want to recapture the joy of a new car.

Is craving the carcinogenic smell of a new car the most irrational piece of car culture? Maybe. But it pales next to the reality that we all hop into our cars every day, knowing that they are contributing to global warming. It is all part of our own denial of the role cars play in our ever-changing world.

TEN

Automotive Maternity
and the Volvo

It is the special prerogative of the poet to stop the speeding automobile within the frame of the page, and thereby enable the public to focus on its debits and credits as an instrument of progress.

—Laurence Goldstein, The Automobile
and American Poetry

In our car crazy society, turning sixteen and getting a driver's license has often been the launching pad for adulthood.

"The most important puberty rite in the United States," Charles L. Sanford says in *The Automobile in American Culture*, "occurs when the young man or woman passes the driving examination, presses down the accelerator, and feels an answering surge of power. It's as if—as some highway poet has written—wolves howled from extinct caves in the bloodstream."[1]

Jacob turned sixteen at Children's Hospital in Boston. We had talked about this birthday for many years. I had promised him a fire-engine red

Jeep as his first car. I thought that it fit my active boy, who loved to play sports and hike.

The carefree spirit of today's Jeep, however, belies its origins as a military vehicle commissioned on the brink of World War II. Made to US Army specifications, the Jeep was designed in seventy-five days with the first prototype delivered on Armistice Day 1940.

US Army chief of staff General George C. Marshall described the Jeep as "America's greatest contribution to modern warfare." As Scripps Howard war reporter Ernie Pyle once said, "It did everything. It went everywhere. Was as faithful as a dog, as strong as a mule, and as agile as a goat. It constantly carried twice what it was designed for and kept going."[2]

That is likely why the spirited little open-air vehicle has been voted the most "patriotic" of brands for twenty years straight, beating out Disney and Amazon for the top spot in 2022.[3]

Often pictured at the beach or on a ski trip in the mountains, the modern Jeep had a carefree vibe with open sides and a roll bar on top. That's how I saw Jacob in one. I wanted him to be out enjoying his time with friends.

But I hadn't counted on his illness. Instead of buying him a Jeep and watching him happily drive off into his future, I wrapped up a two-inch toy model and handed it to him for his birthday. He rolled it across his hospital bed before falling back asleep.

It would be well over a year before Jacob learned to drive. He qualified under a special program in Massachusetts for children who could not be taught in the classroom, so we were able to hire an instructor to come to the house. A kind middle-aged woman helped him with the test and took him out to practice his road skills.

Jacob's early driving was nail-biting. He'd lost his natural fear of death after so many days and weeks battling for his life at Children's Hospital. He had the swift moves of a soccer player seeing small spaces where he could move quickly without thinking carefully about the cars around him.

"You just look for the hole, Mom," he said, breezily zooming across multiple lanes before exiting the highway.

Driving is one of the most dangerous things we do, and even though automakers have fought tooth and nail against almost every safety device that is now standard—including seat belts and airbags—automakers have somehow managed to make safety a differentiator for their brands.[4]

Having almost lost my son to childhood leukemia and the rigors of treatment, I wanted him as safe as possible in the car. So, although I had promised Jacob a sporty Jeep, we would, as a family, settle on giving him our old 2007 Volvo station wagon for the simple reason that I believed it to be a safer choice.

＊

The Volvo station wagon was de rigueur when we first moved to Cambridge, Massachusetts, in the early 1990s. With a Hudson's Bay blanket covered in golden retriever hair tossed in the back and a beach sticker from the summer house on Cape Cod plastered on the side window, the car reeked of old money and was part and parcel of a proper pedigree.

Newer versions, like our 2007 model, were better known for safety and were thought of as a solid, sensible purchase. Volvo repeatedly advertised itself to young families as a virtuous and pragmatic choice. Consider the ad Volvo famously ran in women's magazines in 1991, featuring a sonogram of a fetus with its controversial tagline: "Is Something Inside Telling You to Buy a Volvo?"

With its clear reference to the always heated abortion debate, the ad inflamed both pro-choice and anti-abortion activists while co-opting one of the world's most sacred images. In short, the ad was successful because it pissed off just about everybody and garnered lots of attention for the company.

The ad also reinforced long-standing beliefs about automotive culture: that women favored safety over speed and performance, that women prized utility in cars rather than pizazz, and that the interior of the car was itself womblike, a special domestic space that women were responsible for keeping clean.

Yep, "automotive maternity" is a thing: a verified idea that is part of the distorted and fascinating package of ideas that make up car culture. There was even a short-lived TV show that aired for a single season in 1965, *My Mother the Car*, in which the lead character's mother, played by Dick Van Dyke, is reborn as a car.

So prevalent was this concept that one of the most admired artists of the twentieth century, Pablo Picasso, poked fun at the idea of automotive maternity in his 1951 assembled sculpture *Baboon and Young*. Picasso used two of his son's small toy cars—one upright, one upside-down, to form the face of a mother baboon holding a baby.[5] An automobile spring was used for the mother baboon's tail. Acquired by Peggy Guggenheim and currently in the permanent collection of the Museum of Modern Art, the sculpture was displayed in 2021 in New York at the Museum of Modern Art exhibition *Automania*, next to a Volkswagen Beetle whose front grill echoes the lines of the mother baboon's face.[6]

The sculpture is entirely evocative of the controversial yet powerful idea that cars offer themselves up as a source of motherly love. As Dr. Ernest Dichter, a psychiatrist and well-known Detroit market research analyst, told *Motor Trend* magazine in 1967, climbing into a car is like a return to the womb: "We feel protected and blissfully secure."[7]

The car as a womb has all sorts of ramifications both real and metaphorical. Despite the fact both men and women prize safety for themselves and their children, automotive advertisements historically pitched this idea to women.

As historian Katherine J. Parkin explains in *Women at the Wheel: A Century of Buying, Driving, and Fixing Cars*, "When automotive advertisers consistently marketed children's safety in the car to women, they acted on a sentiment embraced by the industry and verbalized in the twenty-first century by French marketing strategist and cultural anthropologist, G. Clotaire Rapaille, that 'women care about car interiors' because they're 'programmed to create life' inside their wombs.'"[8]

Encoded within these messages was the idea that the internal space of the car belonged specifically to women. That it was both a roving womb and a mobile domestic space akin to a living room. "It is fairly safe to assume that not nearly so much thought and effort would have gone into such considerations as seat cushioning, noise prevention, ventilation and temperature insulation but for the realization that these matters are important to women," read a 1932 ad for Fischer Auto Body titled "A Coach for Cinderella." "Certainly, body-styling, the uses of color, the attention given to upholstery, interior trim, fittings, and equipment conveniences have had the demands of feminine censorship as their standard."[9]

This idea that women were especially capable of decorating a car while not all that adept at driving one is consistent throughout both advertisements and popular culture and was a strategy often used to sell cars to women without offending men. He picks the engine; she picks the paint job.

In 1955, for example, to get the "feminine perspective" on car design, several young women were recruited from Pratt's Institute of Industrial Design by Harley Earl, who was then vice president of styling at General Motors. The legendary design maven was known for his lavish body designs, which included exaggerated tail fins and headlights and massive chrome bumpers. His "Y Job," a car he personally designed and that was manufactured solely for his use by General Motors and is currently on display at the General Motors Heritage Center in Sterling Heights, Michigan.

Earl, who was singularly powerful at General Motors, called his team of female designers the "Damsels of Design." They were featured in lavish car shows and photo shoots and functioned as escorts for visiting dignitaries. A press release from General Motors at the time explained that they had been hired to figure out what appealed to women. "Not too many years ago, the woman's influence on automobiles was limited to a stern voice from the back seat. Today, besides sharing the driver's seat with men, women cast the deciding vote in the purchase of seven out of ten cars."[10]

Earl reportedly told the newly hired designers to think of the car as they would a house.[11] The women were assigned to different divisions, including Pontiac, Buick, Oldsmobile, and Chevy. Even though the women made roughly the same amount of money as the men, their influence was limited. The women were tasked with designing interiors but never allowed to work on exteriors, as that was seen as a purely male domain.

More a publicity stunt than a testament to equality, the women's designs, or "concepts," were touted in "feminine" car shows that were held in 1957 and 1958 in the rotunda at General Motors headquarters.[12] The show, titled *Spring Fashion Festival of Women-Designed Cars*, was open only to women members of the press and included ninety singing canaries and large pots of hyacinths to scent the air.

The stunning cars, whose interiors the women had designed, had special places for children's toys, sets of matching luggage, umbrellas, and makeup mirrors. Some had special compartments to store cameras, picnic food, and utensils. The Damsels of Design were reported to have put in the first retractable seat belt and the first child lock on the back doors that was operable from the front. But the Damsels' tenure was short-lived. Earl retired in 1958, just three years after they were hired. His successor was unequivocal: "No woman stylist will ever be photographed standing next to one of my cars."[13] Most of the women were gone by the end of the decade, but the idea that the car interior was somehow a feminine domain continued.

"Promises of comfort and safety also assured consumers that the car's interior would be easy—for women—to clean," according to Parkin. A *Ford Times* magazine article, for example, described a woman driver as a "hostess," who ensured that the family car was "as clean as her living room."[14]

The idea of the female driver as "hostess" has played out in myriad ways, but one clear way that it can be seen is in the rise of the cupholder in cars. The 2019 Subaru Ascent has nineteen of them. That's more than any mass-market vehicle ever produced, amounting to almost two and a half cupholders for each passenger. There's room for a Starbucks skinny latte, an unnaturally colored Big Gulp, a Yeti Rambler, and juice boxes galore. So many cupholders, in fact, that the *Wall Street Journal* recently declared: "We are approaching peak cupholder."[15]

Long commutes and active, harried schedules have all contributed to the rise of the cupholder. But one academic links the desire to travel with warm beverages back to our earliest needs for warmth and succor. Anthropologist G. Clotaire Rapaille claims that sipping a warm liquid while speeding down the highway is an act akin to reaching for a mother's breast. It is, Rapaille says, a necessary component for our view of the car as safe. "What was the key element of safety when you were a child?" he asks. "It was that your mother fed you, and there was warm liquid. That's why cupholders are absolutely crucial."[16]

Well maybe, but as late as 1989 *US News and World Report* was still calling "crannies for drinking cups" an unnecessary "future frill."[17] But that kind of rhetoric denied the necessity of cupholders.[18] After all, they do keep you from spilling things while traveling at sixty miles an hour.

Consider the 1992 case of Stella Liebeck, who purchased a fifty-nine-cent cup of coffee at a McDonald's near Albuquerque, New Mexico. Her grandson, who was driving his Ford Probe, pulled over to let her put cream and sugar in the cup. The Probe did not have cupholders, so she put the cup between her legs and spilled the coffee, which had been heated to over 180 degrees, on her cotton sweatpants, which absorbed the scalding liquid, severely burning her thighs and groin area. She sued McDonald's, arguing that the coffee was too hot to consume and a hazard. In the first instance, McDonald's argued that the coffee needed to be that hot because drivers wanted to drive before drinking it. Later, it would become clear that most drivers consumed the coffee immediately. Liebeck won a judgment of close to $3 million but later settled for an amount that was reported to be closer to $600,000.[19]

Pleasure (1916). The Baker Electric was originally sold to women so they could enjoy idyllic nature-filled afternoons with their children. *From the Collections of The Henry Ford.*

A Car for Her, too! (1928). "The time has now come when 'a car for her, too' is a necessity." Chevy got an early edge on marketing to women with its "bigger and better" Chevrolet. *From the GM Heritage Center.*

Woman's Place is in the Ford (1949). After World War II, women's work turned from wartime production to household chores. This ad extols the many ways in which the car had become integral to women's work. From chauffeuring her husband to delivering groceries to nursemaid, a woman's place is behind the wheel. *From the Collections of The Henry Ford.*

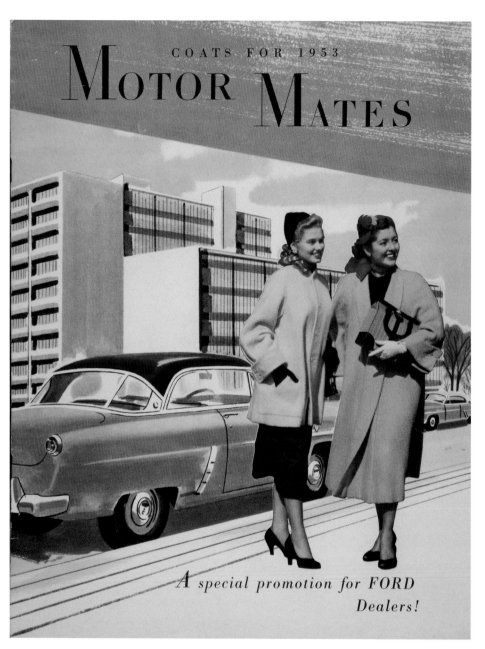

Motor Mates (1953). For a special promotion, Ford created a set of coats and handbags for women drivers using Ford Victoria nylon fabric matching the interiors of their cars. *From the Collections of The Henry Ford.*

Cadillac Sedan de Ville (1956). The link between fashion and automobiles is made clear in this ad shot in the salon of designer Jean Dessès. *From the GM Heritage Center.*

Playroom on Wheels (1958). Peggy Sauer was one of a group of women who worked at GM who were referred to as the "Damsels in Design." The women, hired by the head of GM Design, Harley Earl, were only allowed to work on the interiors of cars. Here is Peggy with a design she created for the interior of a 1958 Oldsmobile that seeks to accommodate a child's toys. *From the GM Heritage Center.*

Mod Dresses to Match Your Camaro (1966). These dresses were made as part of a collection of clothing to match the Camaro. *From the GM Heritage Center.*

Joan Didion (1968). An iconic Hollywood portrait by Julian Wasser of the author in front of her Corvette Stingray came to define Didion's romantic and angsty relationship with the automobile. *Road & Track* called the third generation Corvette that Didion owned: "A perfect symbol of the America that Didion chronicled." *Copyright © Estate of Julian Wasser. Reproduced courtesy of Craig Krull Gallery, Santa Monica.*

The case highlighted the ever-growing need for car interiors to be made hospitable for long commutes and occasional road trips. Form should follow function as every good design student—man or woman—knows. Women care about car design because having your genitals burned by scalding liquids hurts. Just like scalding your child with hot cocoa hurts.

Which brings us back to the car as a protective womb. As late as 2019 *n+1* editor Dan Albert may have summed up the idea as succinctly as anyone in *Are We There Yet? The American Automobile, Past, Present, and Driverless.* "The automobile has become a wandering womb protecting a fetal driver," he writes. "In the event of even a minor crash, much of the car crumples like tin foil to absorb the blow. The vehicle sacrifices itself to save its occupants."[20]

And isn't that just what mothers are supposed to do? Sacrifice themselves for others?

❋

Jacob would eventually become a safe and considerate driver. After finishing high school, he drove that old Volvo station wagon out west to college in California, where he swapped skis for surfboards and skateboards.

I, on the other hand, crashed hard. During his illness I had been like those air bags, inflating around him trying to keep him alive and making sure that the impact of his treatment—although devastating—was not deadly.

He had survived, but I faced an unthinkable void. I had left my professional life with a single email with the subject line DEVASTATING DIAGNOSIS. His treatment had left him immune-impaired, and for the first year I saw almost no one and kept up no professional contacts; I was devoted to him. I wouldn't have done anything differently, but I don't think I realized what I'd done to myself. I was barren.

And then my old Honda Pilot started to go. It was as if it knew that a new day had dawned. Things large and small kept breaking on the car, and the seats ripped, exposing their foam-like interior. I patched them with duct tape until they gave way again.

I kept my Pilot until the engine mounts started to rust through, making the car veer toward the left unexpectedly and vibrate like crazy. The repair costs started to add up, and I could get barely a few hundred dollars for it as a trade in. I knew it was time to let it go. Still, I wept as I grabbed my rosary out of the console and watched my Pilot roll away on the back of a flatbed truck. I didn't know if I was mourning the end of the car or the end of my son's childhood. That car was my church, and it had held every prayer and aspiration I had for my family. But I couldn't repair it and couldn't afford to store it as a token of the past.

As I drove it to the lot to sell it, I cried big ugly tears, which led my husband to ask in an unkind voice: "What is wrong with you?" He never was a car guy, and I knew I couldn't explain it to him.

My husband drives a nondescript, uncomfortable, and perpetually dirty four-door Volvo sedan that I call his "junior executive" car. Truly, if I weren't married to him, I would never get in it. Sitting in the passenger seat, I feel like an egg in an egg carton. Safe, but stuck. Which of course is the downside of opting for a safe and secure path. You don't get hurt, but you don't really get to spread your wings or have much fun.

Luckily, he left the decision of what to get for my next car up to me.

ELEVEN

My Subaru:
Gender and the Lesbian Car

Someone once said that people were never as naked as when they dressed for a party. The same was true when Americans made substantial symbolic, emotional, and financial investments in automobiles. Whether the car was a Model T in the 1910s, a Chevrolet in the late 1920s, a Cadillac or a Volkswagen in the 1950s or an SUV in the 1980s or 1990s, the odds were that its owner at the time was anxious about his or her place in the world and using the car in question—in part—to ask a larger community for some indication of affirmation and respect to allay that anxiety.

—Tom McCarthy, *Auto Mania*

I started shopping for a new car sometime after my son went to college. At first, I tried to get a new Honda Pilot, but over the course of almost a decade the newer models had become engorged, so much so that they no longer fit in my admittedly narrow driveway.

In my heart, I really liked a classic Land Rover, with its boxy style and big windows. Range Rovers—an iconic British brand—were a favorite of Queen Elizabeth II for most of her seventy-year reign. Many of the vehicles she drove were customized to suit her needs and styles. One of the queen's Rovers, an L322, had a custom hood ornament that featured one of her Labrador hunting dogs with a pheasant in his mouth.

Throughout her life and reign, Queen Elizabeth II was both a car lover, a driver with military training, and a mechanic. So much so that when she repaired and maintained ambulances during World War II she was known as "Princess Auto Mechanic."[1]

One of my favorite car stories features the queen. In 1988 she took a reluctant crown prince from Saudi Arabia on a Sunday drive through the rugged grounds of her enormous Balmoral estate in Scotland in her Range Rover.[2]

At the time of the visit, women were not allowed to drive in Saudi Arabia. A retired British diplomat has suggested that the queen took a particularly difficult route at high speed around her estate, which includes several water hazards, to make sure that the prince got the message about women's skill and power behind the wheel. Reportedly, she talked the whole time while the somewhat nervous prince urged her to keep her eyes on the road and hands on the wheel.

Unfortunately, however, I am not a queen, and one of my friends quickly vetoed the idea of my racing around the suburbs in a Land Rover. "No," she said. "You'll look like a rich bitch."

She meant this in the kindest possible way. Drive too nice a car and you are likely to get nicked in the parking lot and challenged at every intersection by someone with an old Nissan that they don't mind banging up to make a personal point about income inequality. Besides, I was tired of driving a big car around town; parallel parking my Pilot was never easy, although I did get pretty good at it.

Further, faced with daunting college tuition payments, I felt that any car I purchased had to last and be economical to run. We think a lot about gas mileage when we think of energy efficiency, but there is a ton of embedded energy in the manufacture of every car. Keeping a car for a longer period of time can be easier on both the pocketbook and the environment.

In the end, I opted for something that was easy to deal with on all levels. One friend had an economical Subaru Forester that she loved and suggested I try it. Priced at under $40,000, the car was both affordable and nondescript in almost every way.

Another friend, also a Subaru driver, was adamant that I opt for the package with the leather seats that heat in winter. She argued that leather seats are easier to clean, and that they don't smell the way fabric ones tend to over time. She looked at me and said, "You know you're going to have to clean it, right?"

I skipped giving my friend a lecture on the historical vestiges of gender and the automobile and just took her advice. She was right; I was going to have to clean it, whether or not I thought that was fair.

So, in the end, far from the kind of sexy relationship some people have with their cars and miles away from my love of the Pilot, I opted for an economical car that I hoped would last and not have smelly seats. I bought a 2015 Subaru Forester.

Once I got it home, I realized that my Subaru had one other attribute that I adore. At just about sixty-eight inches high, it is the first car that seems to really fit me. In other words, it's my size. If I need to put my skis on the roof rack, I can reach it without a problem. If I need to get the Christmas tree down from the roof, also not a problem. Toss the golf clubs in the back, no big deal.

My Subaru is all utility and, although it holds a lot of sporting equipment, the vehicle itself is not that sporty. It has very little kick to it, so I need to time my entry onto the highway carefully, or risk getting smushed.

What it does have is a hatchback design that is perfect for everything, from twenty-pound bags of dirt to groceries. In fact, car journalists call hatchbacks like mine "grocery-getters." Why? Because that's what they were designed for, and that's how they have been marketed to women for almost three decades.

The hatchback was designed for "running around town and shopping," writes historian Chris Lezotte, who documents the evolution of the Chick Car in her article in *Popular Culture* magazine.[3] In 1978, for example, Honda advertised its Civic hatchback in a national campaign as having been designed to hold four upright shopping bags in the back—an attribute that was supposed to appeal to women, who did much of the shopping.

You can see the remnants of that kind of thinking in my Subaru. It does indeed keep four grocery bags perfectly upright in the back. Moreover, its low to the ground design makes it easy to lift the grocery bags without having to bend too far over—as I would have to do if they were in, say, my husband's Volvo sedan.

The fact that the trunk opens automatically when you press a button on your key fob is another benefit designed for women, especially useful for women juggling grocery bags and supervising children while wearing heels in the grocery store parking lot. And that describes an awful lot of us who have worked all day, picked the kids up at school, and then stopped to pick up a few things for dinner.[4]

In that way, my Subaru is a good example of how certain models have come to define a certain subset of the population. Once manufacturers identify an element or attribute that resonates with a certain gender, say the hatchback, they market to them based on the attribute and certain existing social stereotypes. For example, I bought my Subaru because it fit me in terms of both my size and my social standing. It helps me meet my diverse obligations as a caretaker, a gardener, and a golfer and skier.

Once on the road, other people look at the car and the people who drive it and say, "Gee, that's a car for a person who is active and somewhat economical. I'm kind of like that, so I will get that car." In that way, the whole thing becomes a self-reinforcing loop. All of which goes a long way to explaining how cars have become our most gendered and performative technologies.

But the Subaru comes with a twist—a unique identity that makes it stand out in car culture. Subarus are cars that are often identified with lesbians. Seeking to exploit a nascent market niche, Subaru was the first automobile company in the United States to focus a robust sales pitch on gays, with a particular focus on lesbians. It all began in 1994 when marketers at Subaru were seeking to grow the brand. As they conducted focus groups around the country, it became clear that lesbians were already buying Subarus in large numbers. The car was economical and had all-wheel drive, making it dependable for women who might be living alone and often worked essential jobs, such as nurses or emergency technicians, that required them to be on the road in all weather.

At one focus group held in Northampton, a small town in central Massachusetts, all the women who showed up seemed to know each other, and all of them were lesbians. Tim Bennett, who worked for the brand at the time, says that further research showed that lesbians indexed five times higher than other consumers in both purchase and interest in the brand. (Gay men were not as interested in the brand at the time, preferring Jeep Wranglers or BMWs.)

To capture the attention of lesbians the company began to appeal to them using coded words and slogans with double entendres in what one executive called a series of "winks and nods." Cars in the ads, for example, had plates that read PTOWN, a reference to a popular vacation spot for gays in Massachusetts. Other ads featured the tagline "GET OUT and STAY OUT," which, of course, could mean either get outside and stay outside, or come out of the closet and stay out of the closet.[5]

The ads were groundbreaking. For the first time ever Subaru's overt efforts to appeal to lesbians allowed "some women to affirm their gender and sexual identity with the purchase of their cars," says historian Katherine

Parkin.[6] So much so that, as Parkin muses, "One woman maintained in a 2007 *Los Angeles Times* op-ed that a man lost interest in her after one date because he discovered she drove a Subaru Outback and he suspected she was a lesbian."[7]

According to Parkin, the message of these ads is important, because advertising does not just sell goods: it sells meaning. And especially in this case, the ads confer legitimacy. "In giving value to objects," James B. Twichell argues in *Adcult, USA*, "advertising gives value to our lives."[8]

❋

There is not a tremendous amount of recorded history on cars and the gay community. But a few years ago, the complicated and sometimes twisted story of gender, sexuality, and cars took a decidedly fascinating turn. In 2015, about a decade after the first Subaru ads aimed at lesbians appeared, former Olympic decathlon champion Bruce Jenner came out as Caitlyn in *Vanity Fair*. In photos shot by Annie Leibovitz, a fabulous Caitlyn posed in a red Donna Karan dress in a color-coordinating 2011 Porsche 911 GT3 RS, a car reportedly gifted to Bruce by his ex-wife, Kris Jenner. Without a doubt, Caitlyn looked gorgeous in Bruce's car.

Sadly, however, around the time of her gender transition, Caitlyn had been involved in a fatal accident in California in which a sixty-nine-year-old woman was killed and five others were injured.[9] Caitlyn, who was not injured, was driving the third vehicle in a rear-end pileup on a coastal highway. Caitlyn, who was driving a Cadillac Escalade, hit the car in front of her, sending that car into oncoming traffic, where the driver was hit and killed. Caitlyn was not charged.

Yet she swiftly felt the sting of female driver jokes all the same, such as when comedian Ricky Gervais quipped at the Golden Globes that with her accident Caitlyn had done little to help the reputation of women drivers.[10] It was a cringeworthy joke, one that landed on an enormous pile of misogynistic jokes about women drivers, jokes that have plagued women since the first days of motoring.

Caitlyn herself is no stranger to female driver misogyny. Remarkably, even after transitioning, Caitlyn continues to show some classic patronizing male patterns when it comes to women and cars.

On YouTube, for example, you can watch Caitlyn teach her female business partner how to change a tire.[11] In the video, Caitlyn times her business partner's tire-change efforts and finds them, shall we say, lacking. These efforts, I might add, are made while the business partner holds a coffee cup and tries hard not to get her lavishly manicured hands dirty. There are the requisite jokes about losing your nuts—ha, ha, meaning the ones used to keep the tire on the car, of course. It's everything you need to know about cars and gender in the modern high-speed mix master that is social media.

❋

While much of social media has simply reinforced the gendered stereotypes that surround the car, there is one notable exception. Social media helped pave the way for women to drive in Saudi Arabia, but it was a decidedly rough road.

In her riveting book *Daring to Drive: A Saudi Woman's Awakening*, Manal al-Sharif recounts her part as an activist in the campaign for women's driving rights in Saudi Arabia. Al-Sharif has a degree in computer science and worked for the Saudi oil company ARAMCO in the United States, where she learned to drive and drove legally like all American women, driving her son to school and going to the grocery store.[12]

Yet when she returned home to Saudi Arabia after fleeing an abusive husband, she was unable to drive given the laws of the country at the time. Like many Saudi women without marital support, she was forced to either wait for a relative to drive her or to pay for private drivers for herself and her young son. In her book she tells how, without a male guardian to drive her, she had to walk to doctor's appointments along a busy road, where motorists taunted her with slurs usually reserved for prostitutes.

With growing frustration at not being able to do the things that were necessary for herself or her young son, al-Sharif, along with other Saudi women, participated in an illegal driving demonstration near the capital. They loaded their gleeful videos from behind the wheel to YouTube with the hashtag #Women2Drive.

In her video, al-Sharif looks into her iPhone and makes a plea for women's rights. "We want change in this country," she says. "We are ignorant and illiterate when it comes to driving. You'll find a woman with a PhD, a professor at a college, and she doesn't know how to drive."

Within the first twenty-four hours, seven hundred thousand people viewed al-Sharif's video. The pushback was swift and fierce. Al-Sharif was arrested in the middle of the night and served nine days in a decrepit prison. The charge: Driving While Female. The year was 2011.

Seven years later, the ban was lifted, legalizing driving for women in Saudi Arabia. The crown prince who was once schooled by Queen Elizabeth II eventually became king and finally lifted the ban on women driving—in 2018! That's right, almost one hundred years after Bertha took her first long-distance drive in her husband's Benz, the women of Saudi Arabia were given the right to drive by themselves. They could, and do, take themselves to work and their children to doctor's appointments. It's a triumph for women, but the story of Saudi women's struggle for driving independence is harrowing.

At the time of the announcement, *Vogue Saudi Arabia* posed a Saudi princess in high heels and black leather gloves in the driver seat of a red convertible parked in the "dancing sands" of the Saudi desert. The cover drew almost instant vitriol on Twitter as several women remained imprisoned in that country for earlier driving violations.

In short, they were still serving time for doing what the princess did so glamorously on the cover of *Vogue* and women around the world do every day.[13] Indeed, it wasn't until 2021 that Saudi activist Loujain al-Hathloul, a leader in the campaign for women's right to drive, was released from prison after serving three years of a five-year term.[14]

❋

One final story about my little blue Subaru: In 2022, as I was leaving a small museum next to a high school in a Boston suburb, I packed a few things into my hatchback and started to climb into the driver's seat. That's when I heard two teenage boys talking about my car. As they walked away, one of them pointed at the Forester and said: "Yeah, that's a lesbaru."

TWELVE

The Pious Prius and the Bicycle: A Story about Violence and the Car

Technology does not simply fall from the sky; rather, the aspirations of a society (or a class) combine with technical possibility to inject a bit of culture into the design like a genetic code.
—Wolfgang Sachs, *For the Love of the Automobile*

My son, who never knew my father, has zero interest in car culture. Kids my son's age seem to feel differently about cars, than, say, my brother's generation did. My son didn't take a shop class like my brother, and he never kept a plethora of parts in the yard as we did in the old days. In general, young people today are much more attracted to their phones and their Nintendos than their cars.

The well-known essayist and novelist Tom Wolfe, who wrote about car culture and its potent mix with youth culture in *The Kandy-Kolored Tangerine-Flake Streamline Baby*, has a phrase for this. He says that corporations and large organizations have simply "routinized the cha-risma" that once surrounded the car and in particular young people's love of the car. [1]

That is, the car culture that once existed for my brother and my father, where they could and did lay hands on a car to make it their own, has been corporatized. People don't work on their cars anymore and, for the most part, they don't customize them much beyond a few accessories.

This is not to say that America's youth have totally abandoned car culture either. In 2021 Olivia Rodrigo hit the top of the charts with her debut single, "Drivers License." In the video recorded for the hit title, the eighteen-year-old former Disney star uses her new license to drive a vintage Mercedes by her old boyfriend's house while pining away for him. Later in the video she inexplicably plays the piano in a car lot at night.

It could have been written when I was in high school, and it might have been shot at my father's dealership. Yet despite the success of the song, young people are embracing that transformative act of getting their driver's license later and later.

No longer enamored by car culture, and often connected more by tech-nology than by roadways, young people today get their licenses later and begin driving with less enthusiasm than earlier generations. In 2014 just 24.5 percent of sixteen-year-olds had a license, a 47 percent decrease from 1983, when 46.2 percent did. [2] A far cry from the 1970s, when every kid I knew in high school had their driver's license by the time they turned seventeen, even those without a family car.

But beyond getting their licenses later, some kids are revolting against the act of driving altogether.

In my son's junior year abroad, he studied in Copenhagen and biked daily to class. After he graduated from college in 2018, he moved back to Boston. Jacob was adamant that he would not give up the freedom of the open air. He would ride his bike.

We had been at odds that summer. I didn't want him biking to the lab where he worked on leukemia research as he prepared to apply to medical school. Boston is known for its narrow roads and aggressive drivers; in 2014 the insurance company Allstate rated Massachusetts drivers the worst in the entire country.[3]

I did not feel it was safe for my son to ride his bike into town, especially in the congested area near the Harvard Medical School and many other hospitals. I also hated the idea of him returning to the part of town where he had suffered so much, where we both had our cancer treatments.

I told him not to ride his bike; he did it anyhow. I sat on pins and needles every day waiting for the screech of the garage door that meant he was home. Then I pretended to be surprised to see him.

"How was your day, hon?" I would say in my sweetest fake mom voice.

There were lots of close calls that summer, but the one I remember most clearly came late in the season as he was biking in Cambridge on a busy street near where we live. Cambridge is home to Harvard University and some of the most left-leaning people in the country. The bike lane had ended, and he rightly took the middle of the lane to cross the intersection. As he waited at the red light, a woman in a Toyota Prius drove up so close behind him that she bumped his tire, apparently intent on pushing him out of the way. She was a bully in a three-thousand-pound car.

As he told me later, when he came home shaken, but unhurt, "She flipped me the bird and took off."

I was outraged, but not surprised, by the bullying tactics of a Cambridge lady in a Prius.

The Toyota Prius is an outlier in the long history of the automobile. It is a hybrid—one of the first—meaning that it is powered by both gasoline and a battery to deliver record-breaking fuel efficiency: somewhere near forty miles per gallon. Given that my son was biking in Cambridge, with its long-standing leftist politics, her car was a social signifier. Like the muscle cars my father sold that signified with their loud roar that men were daring and fast and, well, hard, the Prius signified that she was smart and concerned about the environment.

In fact, those who drove a Prius in the early days seemed to lord their virtuousness over others, so much so that people called it a "Pious Prius." In fact, it was the Pious Prius that inspired a well-known South Park episode titled "Smug Alert," in which smog rates are down due to better vehicles, but "smug" rates are up from hybrid drivers who release their toxic "smug" into the air.

The bully driver probably bought her Prius to help her fit into her social class, to help her create a persona that showed her discernment. Outside of left-leaning communities, however, the Prius is a car that mostly gets ridiculed. Consider this dialogue from the movie *The Other Guys*, released in 2010 and starring Will Ferrell and Mark Wahlberg. In the movie, two misfit New York City detectives try to stop a crime wave in a Prius, a car that is characterized as both female and a piece of shit.

For example, Wahlberg says to Ferrell as they cruise around in the Prius: "I feel like we are literally driving around in a vagina."

Later, another detective sizing up the Prius quips: "I didn't know they put Tampons on wheels." After which he mocks its size: "My Suburban crapped out one of those last night."

Surely these are not the perspectives our Cambridge lady considered when she bought her Prius. But cars, even as they define us, seem to have a special ability to provoke derision. Think of the middle-aged balding man with a noticeable tummy getting out of a Corvette at his golf club, a look that is much more desperate than inviting. While he believed he was purchasing youthful sex appeal, onlookers are often simply embarrassed for him. On the road we are at once separate and together in our cars in a truly revealing way.

"Traffic flow is a shared good of an interesting sort," Matthew B. Crawford writes in *Why We Drive: Toward a Philosophy of the Open Road*. "It is a fragile, emergent property of a collective, a state that happens only if everyone is paying attention to the situation and brings a disposition of flexibility to it. At times, it resembles an improvisation among musicians. Urban driving at its best is an experience of civic friendship, an act of trust and solidarity that makes one proud to belong to the human race."[4]

At its worst it can turn violent in ways ranging from bullying a boy on a bike to a road rage incident that turns deadly. It's everything from cutting someone off to firing a handgun at another car while driving on a highway, as someone recently did in Miami.[5]

Part of this is simply the nature of the technology. Everyone who is driving has access to power way beyond their natural abilities and a choice about how to use that power. The new "full self-driving mode" on a Tesla, for example, comes in three modes: chill, average, and assertive.

Moreover, as Crawford writes in *Why We Drive*, the car itself depersonal-izes everyone. You don't see Jane or Joe. You see a whole slew of stereotypes: "Women drivers. Macho drivers. Clueless old people. Heedless youth. Asian drivers. Prius drivers who want you to know they 'buy local.' Meth-head red necks in jacked up pickups. Blow dried douchebags in BMWs. Vindictive fat people in Pontiac Aztecs."[6]

Sometimes we even find ourselves bragging about our own bad behavior when we recount that we cut off some asshole with a Trump sticker. Or boxed out some guy with a Biden sticker.

Beyond everyday aggressive driving and out-and-out road rage, there is evidence that the use of the car as an actual weapon is increasing. Recent incidents include the horrific use of a van by Islamic terrorists on Bastille Day in 2016 in Nice, France, that killed eighty-six people. Or the tragic 2021 killing of six people at a Christmas parade in Waukesha, Wisconsin.[7]

And the disturbing cases of people driving violently into crowds of Black Lives Matter protestors. All told, there were over one hundred interactions between cars and protestors during the protests set off by the killing of George Floyd between May and September 2020.[8] Over forty of these incidents were thought to be malicious, and some thirty-nine people were charged.[9]

And who could forget the moment in 2020 when Trump supporters formed a vehicle parade—dubbed the "Trump Train"—that forced a Biden campaign bus off its intended route in a politicized use of the auto that had never been seen before. The bus was traveling on Interstate 35 in Texas when it was forced to abandon its route, despite repeated calls to authorities for help.[10] When pressed, local authorities said that, while they had investigated the hostile act, there were no explicit violations. A lawsuit is pending.[11]

Says law professor Gregory H. Shill, "Our legal system forgives few acts of violence so readily as those committed with a motor vehicle—even those done on purpose."[12]

And that includes acts of aggression as small as a Prius driver intentionally bumping a kid on a bike and as egregious as the bullying instigated by a bunch of presidential supporters against their political foes carried out on a federal highway, supported by all our tax dollars, in full view of the authorities.

THIRTEEN

COVID-19, Cars, and the Hated Commute

*Machines promise power, mobility, freedom, even a poetic space
that beckons from beyond the too familiar course of things, from
beyond the rush of time and time's sad waste.*
—Leo Marx, *The Machine in the Garden*

In March 2020, as COVID-19 emerged as a global threat and the world shut down, our relationship with the car changed once again. With schools and businesses closed, cars for the most part remained silent in the driveway. Yet despite the fact that fewer people were driving and roads were nearly empty, the car became more, not less, deadly as drivers took advantage of the lack of congestion to reach recklessly high speeds.

Women, for the most part, unless they worked in essential jobs such as in health care or at grocery stores, found themselves at home with their kids and spouse, the car parked idly in the driveway.

Then, as the weeks of isolation turned into months, women discovered that cars were just about the only safe way to gather. As a result, there were drive-by birthday celebrations, drive-by funeral viewings, and drive-by graduations.

In my small suburban town, parents organized a parade to celebrate high school graduates, with fire trucks leading graduates in a parade that traveled every street in town to the loud cheers of residents gathered on front steps.

Starved for camaraderie, women sometimes met for a coffee in an open hatchback with friends, or frantically snuck out to desolate parking lots searching for someplace quiet to Facetime with hospitalized family members miles away.

Women who were precariously balancing work, childcare, and homeschooling took Zoom calls from the car. So much so that Nissan, desiring to meet the moment, developed a concept van that included an entire working office pod that slid out the back, complete with a chair and a desk for a computer. [1]

Yet even as cars became parked containers for our work, our loneliness, and our grief, they also returned to their historic sites for romance. Couples snuck out to the driveway for late-night sex to avoid college kids who had returned to the nest to study and play computer games well through the night. [2]

In addition, car interiors became beacons for those wishing to gain internet stardom as their soundproofing made them readily accessible portable recording stations—phone mounted to the dash—for those wishing to do late-night confessionals about something that was bothering them; it turns out a lot of things were bothering them.

Things were turbulent in my house as well. My son, newly enrolled in medical school, began his studies from his childhood bedroom and attended cadaver dissection lectures via Zoom in the dining room. All three of our cars—two old Volvos and a low-mileage Subaru—sat idly in the driveway. We simply had no need for them. There was no place to go except the grocery store. For the first time in a long time, I heard the birdsong of spring and felt at peace in the backyard without the rumble of trucks and the smell of exhaust.

Seismologists who study the effects of human activities on the earth found that our world got quieter because there were also fewer man-made vibrations from cars and buses. In a study published in *Science* in July 2020, researchers who study the noise made by vibrations—man-made and natural—discovered that the COVID-19 pandemic had unleashed a "wave of quietening." It was, they said, "the longest and most prominent global anthropogenic seismic noise reduction on record."[3]

Experiencing the calm and silence of the pandemic, we all became more fully aware of the costs of constant noise from cars. For example, in 2022 the city of Paris began to install sensors to track traffic noise after discovering that "noise reduces the life expectancy of Parisians by nine months."[4] One report cites noise as the second largest threat to health after air pollution, including higher blood pressure rates and disturbed sleep.

Surprisingly, however, as more people stayed off the roads, the death toll from automobile accidents rose. Researchers at the National Highway Traffic Safety Administration logged almost forty thousand traffic deaths in 2020. That's about 7 percent more than the previous year, when traffic was unaffected by COVID-19.[5]

One theory about the increase in deadly accidents is that the empty roads encouraged high-speed driving. Indeed, car aficionados on social media were quick to post their record speeds. New records were set for the 2,900-mile Cannonball Run, which starts at the Red Ball Garage in New York City and ends at a hotel in Portofino, California. One couple claimed to have made the cross-country trek in twenty-five hours and thirty-nine minutes, with speeds averaging around 117 miles an hour. Their top speed: 175 miles an hour.[6] But experts were quick to point out that empty roads and the lack of police presence were not the only factors in increasing traffic deaths. "Something changed during the pandemic in the way some of us think about driving," warned Emily Stein, president of the Safe Roads Alliance, a nonprofit advocacy group that promotes safe driving.[7]

What changed? Emotional dysregulation. Tired of being imprisoned at home, drivers took to the streets with unusual recklessness and renewed rage. More anger, more alcohol and cannabis, and fewer folks doing sensible things such as wearing seat belts led to record death totals, about 20 percent over pre-pandemic levels.[8]

Those numbers put the United States at the top of the global fatalities list, outstripping other developed nations that also struggled with COVID-19. Some experts theorized that it was our highway system that contributed to our much higher death rate. "The pandemic made more apparent how much American infrastructure contributes to dangerous conditions, in ways that can't be easily explained by other factors," the *New York Times* warned in a front-page article.[9]

Jennifer Homendy, chair of the National Transportation Safety Board, echoed this concern, saying that in the United States, "Motor vehicles are first, highways are first, and everything else is an afterthought."[10]

Death be damned; demand for cars soared during the pandemic as consumers shied away from public transportation. Despite supply chain problems that caused record prices, some consumers willingly paid thousands of dollars over the manufacturer's suggested price just to secure a vehicle. And dealers, ever ready to take advantage of the national crisis, demanded those inflated payments up front to make a deal.

The irony was not lost on headline writers and journalists alike, who saw demand for cars skyrocket even as drivers had fewer places to go. One man stated that he drove over 330 miles to avoid extra charges that pushed the price of a Ford F-150 pickup truck $5,000 over the sticker price. "The rules have changed so dramatically," he told the *Wall Street Journal*. "The dealer's position is 'This is kind of a take-it-or-leave-it proposition.'"[11]

❈

Remarkably, as vaccinations became widely available and the concerns about COVID-19 receded, people did not return to work. It wasn't that they hated their jobs or their bosses, although some did. Rather, what they hated was the commute. In comment after comment on news articles, workers cited the commute as time-wasting, stressful, irritating, and exhausting. They might have missed their coworkers, but they didn't miss the hassle of getting to work.

The virus may have accelerated the work-from-home trend, but it was the car and the traffic it produces that cemented it. Is this good for women? I don't think so.

Women have always had to work a second shift when they came home from the office, doing a disproportionate share of domestic duties in the household. [12] My concern is that women who are now primarily working from home will simply add on more and more chores between Zoom calls and Slack messages. Moreover, those same women who are doing the extra housework are more likely than men to be penalized by their employers for working at home.

One article in the *Economist* cited a pre-pandemic study that showed that women suffer more from "flexibility stigma" than men. Meaning that they are likely to be paid less and promoted less often than male employees who work from home. [13]

As a result, women are caught in just another double bind: fight the traffic on the way to work or stay at home, do more housework than ever, and be undervalued.

Recognize also that in no scenario do women avoid being in the car, whether it is commuting to work or ferrying kids to school, piano lessons, and soccer. In the end, the car—the very thing that was supposed to set them free—has enslaved them.

FOURTEEN

Paying the Pink Tax: The High Cost of Driving While Female

The purpose of art is to lay bare the questions which have been hidden by the answers.

—James Baldwin

One thing that my research for this book has shown is that women pay mightily for our cars. We pay more than men when we purchase cars, we pay more than men when we get our cars repaired, and we pay more than men in terms of bodily harm when we are involved in car crashes.

While men are far more likely to be involved in car crashes than women, women are much more likely to be seriously injured in a crash. "When a woman is involved in a car crash, she is 47 percent more likely to be seriously injured than a man, and 71 percent more likely to be moderately injured, even when researchers control for factors such as height, weight, seat-belt usage, and crash intensity," according to Caroline Criado Perez in *Invisible Women: Data Bias in a World Designed for Men.* "She is also 17%

more likely to die." Perez contends that this "all has to do with how the car is designed—and for whom."[1]

As I mentioned earlier in the book, the car was born in a masculine manger, and it is fair to say there were few to no female engineers involved in the creation of the modern automobile. Indeed, as early as 1942 one safety expert accused automobile manufacturers of creating a hazard through their failure to "consider drivers who vary considerably from the norm in stature," and he pointed out that "women and short people in particular had difficulty reaching the controls and seeing out of the automobiles."[2]

Even cars that are "safe" for most people can be "off" for others. For example, I have trouble driving my husband's Volvo even though, for all intents and purposes, it is a very "safe" car. The A pillar—the front bar in the car that surrounds the windshield and begins to frame the doors—is big and sturdy. Presumably, the fact that the car is well engineered and well made makes it safer. Certainly, if you are imagining a rollover that A pillar is a plus. But I am a short woman, at five foot three inches, with short legs. Like lots of women, I sit too far forward when I drive. That means the A pillar creates a blind spot for me when I turn left; in one instance shielding a pedestrian in a crosswalk from view. Therefore, the very thing that makes my husband's car safe for him makes it less safe for me. As a result, I must constantly remind myself of the blind spot when I drive my husband's car.

This kind of complexity makes every engineering decision and regulatory decree fraught with difficulties. Car companies employ complex global supply chains to design multiple products for diverse international markets. However, the vast size of the US market means that most global manufacturers seek to design automobiles that are competitive in the United States. As a result, US regulatory standards can have an outsized effect on the global market.

To determine safety ratings and to do a whole lot of safety testing within the design process, regulators and car companies use crash dummies to simulate all sorts of terrible crashes—head-on, rear-end, and side collisions.

The first crash dummy was made in 1949 and used to test the ejection seats made for male fighter pilots. Nicknamed Sierra Sam, the original dummy was roughly the size and weight of the average man at the time: six feet tall and 170 pounds. Sierra Sam and other mannequins like him, more properly called anthropomorphic testing devices, were later adopted by the automotive industry, where they were put to the test, among other things, by being strapped to sleds and propelled into walls at high speed.[3]

Today's dummies cost as much as one million dollars, and they are loaded with sensors that send detailed information to engineers and technicians upon impact. Federal regulators didn't begin using a female dummy until 2003, and while automakers now maintain a roster of different dummies, including ones that approximate a pregnant woman and ones for small children, the official mannequins that the government uses for federal safety testing have drawn widespread criticism for failing to accurately represent women.

The standard dummies used by the National Highway Traffic Safety Administration to provide vehicle safety ratings for consumers include two versions for women. One version weighs 108 pounds and is four feet eleven inches tall. The second is the same height but weighs 97 pounds.[4]

These official dummies are just "a scaled-down version of a male dummy," according to Consumer Reports. In addition, they "represent only the smallest 5 percent of women by the standards of the mid-1970s. So small, that they can work double-duty as a 12- or 13-year-old child."[5]

Moreover, according to *Safety+Health* magazine, these scaled-down versions of the average, midsize male fail to accurately reflect "female geometry, muscle and ligament strength, spinal alignment, and dynamic responses to trauma."[6]

That's a problem because women have less musculature in the neck and shoulders and other anatomical differences, which means that they absorb impact differently—specifically, but not limited to, a greater incidence of whiplash.[7] According to Consumer Reports, "Females are up to three times more likely to suffer whiplash injuries than males, but real-world crash data shows many vehicle seats that have been specifically designed to prevent whiplash injuries are actually less likely to help female occupants."[8]

Results from studies using the poorly designed crash dummies have likely been baked into dozens of design decisions, many of which are prejudiced against or at the very least not representative of women. "Because automotive design is directly influenced by the results of safety testing, any bias in the way cars are crash-tested translates into the way cars are manufactured," Keith Barry writes. "So, if safety tests don't prioritize female occupants, carmakers won't necessarily make changes to better protect them."[9]

Happily, there is some evidence that cars are getting safer for women, particularly if you are fortunate enough to be able to afford a newer car. Recent research by the National Transportation Safety Board, for example, shows that women involved in accidents in vehicles from the model years 2015–2020 had a fatality risk that is just 2.9 percent higher than males.[10] However, because the average car on the road is almost twelve years old, that leaves lots of women at increased risk.[11]

Faced with pressure from over sixty-five female members of the US House of Representatives, the National Highway Traffic Safety Administration has set out to change federal safety testing patterns, with some hope of getting newer and better female test mannequins into service sometime in the future.

In the meantime, gender bias in automobile design continues to wreak havoc on women's bodies. Consider Maria Weston Kuhn's story. Writing in *Ms.* magazine in 2022, Kuhn recounts a terrible accident in which she and her mother were hit head-on. Ms. Kuhn suffered numerous injuries, some of which she thinks better safety testing might have prevented. "The seatbelt, for example, was designed to stop a man's forward momentum by catching his rigid hip bones," she writes. "For me, it didn't. It slid above my hips, pinned my intestine against my spine and ruptured it."[12]

As Kuhn reminds us, car crashes can be both unexpected and unintentional, but in this case, she writes, "It is no accident that women are more likely to be injured and killed on U.S. roads."[13]

Particularly because the sheer size of today's vehicles is daunting. Traffic safety expert Gregory H. Shill calls it "the engorgement of the American vehicle."[14] As a small woman, I can stand in front of a Ford 150 pickup truck and be entirely invisible to someone in the driver's seat. Try it for yourself the next time you are in the parking lot at the grocery store. Then envision a kid on a bike trying to cross in front of it.

As Jennifer Homendy, head of the National Transportation Safety Board, told Inside Climate News in 2023: "I'm concerned about the increased risk of severe injury and death for all road users from heavier curb weights and increasing size, power, and performance of vehicles on our roads, including electric vehicles."[15]

How did we get to a point where the size of a vehicle can so outstrip its functionality? Part of this is a basic need to feel safe in an unsafe world. Some of this comes down to macho showmanship—as in "my pickup truck is bigger than yours." But that alone doesn't explain it.

Perhaps one of the most important parts of this has to do with profits. Car companies manufacture and promote bigger cars to make more money on a vehicle whose fixed costs don't vary that much as the automobile gets bigger. Parents want to protect their families, and everyone wants to feel safer on the road, so people buy the bigger cars. To be clear, cars are much bigger than ever before. A Ford F-150 truck can be eighty inches tall and weigh as much as seven thousand pounds.

"The quest to be safe has led to its own arms race where bigger and safer cars now dominate on the road becoming a menace to those who either don't want a behemoth vehicle or can't afford one," Jamie Kitman of *Car and Driver* warns. Ironically, this quest for safety has made the roads more dangerous for everyone.

※

Beyond paying with our bodies, women also pay more in dollars. One reason that I don't own the cars I drive is because I've never felt that I could get a good deal at the dealership. As I mentioned earlier, I've always had my husband, a litigator, negotiate for and purchase our cars.

The data backs up my concerns.[16] Women pay roughly $7,800 more for a car then men do.[17] On average, they pay several hundred dollars more in purchase price and about $25 more per repair. Experts call it the pink tax.

Recently, for example, I paid just over $80 for an oil change on my low-mileage Subaru. The bill came with $35 for parts and supplies. That made no sense to me since I didn't replace any parts. When I asked about the charge, the service manager sheepishly told me that I had been charged for the rags they used. He later removed the charge.

Cars are a double-edged sword for women. Women need cars to work and take care of children, but especially for poor women, the increased prices of automobiles, car repairs, and the high cost of auto loans—about 8 percent in 2023—make them near impossible to maintain. Not having a vehicle, however, limits where poor women can live, and especially with housing shortages, that's a tough position to be in.

"Subprime lending and longer loan terms have resulted in a doubling of auto loan debt over the last decade and a notable surge in the number of drivers who are 'upside down'—owing more money than their cars are worth," according to Andrew Ross and Julie Livingston. "Championed as a kind of liberation, cars—all but mandatory in most parts of the country—have become turbo-boosted engines of inequality, immobilizing their owners with debt, increasing their exposure to hostile law enforcement, and in general accelerating the forces that drive apart haves and have-nots."[18]

Holly Barchie's family's descent into homelessness was chronicled in the *Boston Globe* in late 2022.[19] She later added to her story on a GoFundMe page in which she pleads for a "reliable vehicle" so she can find a home and a job.[20] Barchie has four children, and her husband works as a roofer.

My next goal is to find a reliable vehicle so we don't have to say, "no we cannot move there because my husband will not have a ride to work" or "it is to far away from the few people we have that helps us make sure we have good food on the table and takes us to all dr appointments! IF ANYBODY KNOWS A RELIABLE CAR DEALERSHIP THAT WILL MAKE SURE WE CAN GET INTO A VEHICLE THAT IS RELIABLE, THAT WILL ALSO LAST US AND ALSO WILL BE BIG ENOUGH FOR MY ENTIRE FAMILY COULD YOU POINT THEM IN MY DIRECTION? I AM NERVOUS TO JUST GO ANYWHERE AND END UP WITH A VEHICLE THAT NEEDS A TON OF WORK OR ONE THAT BREAKS DOWN ALL THE TIME AS I WILL NEED IT TO GET MY HUSBAND TO AND FROM WORK, THE CHILDREN TO AND FROM SCHOOL AND DAYCARE OR THE BABYSITTERS SO I CAN START WORK THE DAY AFTER WE MOVE INTO OUR NEW HOME!! THANK YOU ALL AGAIN AND JUST KNOW EVEN THOUGH I HAVE NEVER MEET NOR SPOKE TO ANY OF YOU EVER, WE TRULY LIVE EACH AND EVERYONE OF YOU!! [*sic throughout*]

As Barchie's story shows, there are complex competing car narratives for women. The imagined, romanticized narrative that includes convertibles and couture dresses, and the other, more painful, reality that includes high costs, debt, and danger.

FIFTEEN

Back to the Future in the Electric "Ladies'" Car

Futurism is a genre of mythmaking that seeks to generate a feeling of inevitability around some desired outcomes, a picture that is offered as though it were a prediction. This is a good way to attract investment. . . One must accept the future rather than "cling to the past."

—Matthew Crawford, *Why We Drive*

It's hard to imagine this now, but in its infancy the car was pitched to consumers as a way to connect with the past. Bristling at train schedules and rigid streetcar routes, those who purchased the first automobiles longed for a way to reestablish their independence, travel their own paths, and reach their destinations at a time of their own choosing. As cultural historian Warren James Belasco has said of the early automobile, "The future came disguised as the past."[1]

Today, history is repeating itself as the car of the future is deeply rooted in our collective automotive past. Our current best choice to replace our dependence on the gas-powered internal combustion engine is an electric car (EV), a vehicle first offered in the early 1900s to fancy ladies such as Henry Ford's wife, Clara.

In 2022 about 15 percent of all new car sales globally were electric, with almost every major manufacturer either producing or planning to produce an electric or a hybrid vehicle. Some states have even mandated that new car sales in the future be limited to only EVs or hybrids.[2] California governor Gavin Newsom, for example, has issued a directive that requires all new vehicles sold in the state by 2035 to be electric or plug-in hybrid electrics (PHEVs).[3] Given the vast size of the market in California, where there are literally more cars than people, that directive alone could revamp the industry.

Storied American brands like Cadillac have now engineered their final models featuring combustion engines,[4] and even the Dodge Charger, the fast, lightweight car that my father sold in multiple neon colors to veterans back from Vietnam, is coming out with an electric version, one whose horsepower can be adjusted by downloading and unlocking new settings via Wi-Fi.[5]

The cars of my youth might have been about sex and power and options like wheel trim or bucket seats, but today the conversation that surrounds cars is both antiseptic and futuristic. Experts in the field talk about personal mobility needs within a highly diverse transportation system, one that is sustainable or at least attempting some sort of sustainability. Tesla's stated mission is to "accelerate the advent of sustainable transport."[6] The newly announced brand Afeela, a joint venture between Honda and Sony, is said to have been so named because sensing and feeling are such a big part of the imminent new driving experience.[7]

All of this touchy-feely futuristic do-gooding comes with a certain nostalgic nod to our past. The new Dodge Charger EV keeps the shape if not the inner workings of a previous, more optimistic and definitely more reckless, time. It's a far stretch from the days when men went fast and hard on the stick shift and women were told to match their lipsticks to their Mustangs. If only our car choices were that simple today.

True to the origins of the industry, the charismatic man who has reignited the future of the electric is about as ornery and odd as the man who started it: Henry Ford. Elon Musk used his earnings from the sale of PayPal to take over Tesla. And while he didn't have a lot to do with the origins of the car, he did see a way into the future that others didn't. That is, he figured out how to make the electric vehicle appeal to a certain kind of man. One who wanted power but also wanted to show he was forward-thinking, tech savvy, and concerned about the future of the planet. It's not hard to imagine a fleece-vest-wearing tech bro sliding into his Tesla with its enormous screen attached to the dash. Musk swept the short history of Clara Ford's "ladies car" and the emasculating Toyota Prius under the front seats, and men took to the Tesla faster than the thing can go from zero to sixty.

Women, however, have been much slower to warm to the electric car. Even though in surveys women profess more concern about environmental issues than men, they've yet to fully embrace the electric vehicle. According to data from Edmunds via *USA Today*, 69 percent of Tesla vehicles in the first quarter of 2019 were registered by men versus only 31 percent by women.[8]

What's holding women back? In the first instance, automakers have been slow to offer family-friendly vehicle designs, such as wagons and SUVs. Price is another deterrent, as are concerns about installing charging equipment at home. Most often, however, it is the fear of being stuck on the road. *Range* is the common word used for how far you can go on a battery charge. And like miles per gallon, range can vary based on weather, weight, and road conditions. Women, who have had their fair share of danger on the road and often bring children along for the ride, don't want the added anxiety that range brings.

Making matters more complicated, charging is not yet standardized; until quite recently that meant that you needed a Tesla charger to charge a Tesla and a different charger to charge a Chevy Bolt.[9] Although that is beginning to change with more and more companies adopting a unified North American Standard.

In other good news on the charging front, seven auto manufacturers, including General Motors and BMW, have recently announced a new initiative to build their own set of thirty thousand supercharger stations. Although that still may not be enough. A recent report from the Electric Vehicle Infrastructure Coordinating Council, for example, estimates that Massachusetts alone will need ten thousand fast-charging connections to meet demand. That's up from the current 651.[10]

"Women don't want to sit in a dark parking lot waiting for their car to charge," Vermont resident Caroline Gambell told the *New York Times*. "Range anxiety is real. If you are trying to get stuff done, and you have kids in the back, the last thing you need is, 'Is my car going to get there?'"[11]

In January 2023, for example, two reporters from the *Boston Globe* took a Kia and a Tesla on a trip around New England. There were better charging choices for the Tesla, including a "gleaming row of 17 superchargers" outside a supermarket. But the Kia driver suffered. While the Tesla charged in about fifteen minutes, the Kia had to wait in a Walmart parking lot where one of the four chargers was broken and another was deficient. It took just over an hour to go from 18 to 80 percent battery strength.[12]

In the meantime, horror stories abound. In one nightmare scenario for EV drivers, all the chargers at all six rest stops on the Massachusetts Turnpike were broken on Memorial Day Weekend in 2023.[13]

Others report that some EV drivers simply park at a charger and then head off, leaving others waiting in line behind an empty fully charged car. More infuriating, drivers of gasoline-powered cars have taken to parking in charging spots to show their disdain for the expensive new cars—an act referred to as "icing"—in reference to the internal combustion engine. Indeed, one car critic has written that working and easily accessible charging stations are "so rare they might as well be folklore."[14]

Commenting on a story in the *New York Times*, a woman named Sarah lays out the dilemma perfectly.

> In a family emergency, I was forced to drive a Tesla home from Boston to New York a week ago. Having never driven a Tesla before, I had no idea when I got in it how to even make it move. How to reverse. How to drive. How to park. How to work the lights. How to work the doors.
>
> There were a lot of really freaky things that happened. I opened the door to take a parking ticket and the car suddenly turned off—seat rolling back all the way to the back seat with a line of traffic behind me. I had to figure out how to start the car again, and literally had no idea how to.
>
> Not only that, but I left Boston with 80% charge, and I had to stop an hour and half into the trip because I had dropped so much on charge that I was perilously close to running out of charge. I arrived to a super charging destination with 7% charge halfway home. I was extraordinarily tense, afraid that I would stall in the middle of the road and have no way to charge, watching the charge tick down. I mean on paper it looks good, but in reality, when you're driving down the road and the charging station is 25 miles away from where you need it and you're on 7% there is a likelihood you're going to stall in the middle of the road by yourself and you can't call a truck to bring you gas. But what you can call is a truck that runs on gas to bring you electricity.
>
> These conversations are very intellectual until you yourself are in one of these cars and you realize how unbelievably helpless you are. [15]

Women like Sarah will undoubtedly adapt to the challenges that new EVs present just as they have to the many other challenges that have come from technologic and automotive change. And, like any new technology, it will improve over time as more and more consumers adopt the EV. As it stands, the EV is our best option to combat global warming, and the Biden administration and many large manufacturers are behind it in a big way.

The Biden administration has set aside $5 billion in its new infrastructure bill to help build out the charging infrastructure nationwide. While in Massachusetts, the state's Department of Public Utilities has approved a $400 million plan to add more high-speed public charging stations. To pay for the charging stations, the legislature will allow utility companies to pass the expense of adding up to forty thousand new high-speed chargers to residents. That's great if you have the money to upgrade to an EV, but it stinks if you're a woman with little interest in a new technology and just hoping your 2015 gas-powered vehicle will start and will be able to get you to work in the morning. With utility prices already skyrocketing in my state, there is the very real possibility that a single mom somewhere is subsidizing the charging of an electric vehicle that is the second or third car in some guy's garage.

Fears of the batteries catching fire is another negative, as battery fires can last up to four hours, as one did in my town in Massachusetts. In 2020 General Motors recalled nearly sixty-nine thousand Chevy Bolts as fire risks, warning consumers not to park them in their home garages.[16] Says Pulitzer Prize–winning car critic Dan Neil of the *Wall Street Journal*, "The Bolt fires absolutely floored me."[17]

The smallish Chevy Bolt was an early favorite to be a female-friendly EV and Chevy has since come out with a safer battery setup and is also offering a free standard level-2 charging station that can be installed in your home. While that's a plus for women who own their own single-family home, it is not clear how this will work for women who live in an apartment complex.

The *Wall Street Journal* reports that "the federal government will soon start to release the first $700 million of $2.5 billion planned for discretionary grants that would place chargers at places such as schools, stores and apartments."[18] That's a start anyways.

A Chevy ad that ran in early 2023 wished a Happy New Year to "EVeryone, EVerywhere," and featured a lineup of electric vehicles fronted by three women of color in cheery sundresses, a throwback to an earlier Chevy Camaro campaign where three White women with bouffant hairstyles in wild print dresses adorned the speedy racer of my youth. In a nod to the concerns women have about access to service on a car with as-yet somewhat unproven technologies and a record number of recalls, Chevy's current ad touts its vast dealer network and astoundingly claims that for 90 percent of Americans there is a Chevy dealership within ten miles of their front door.[19]

To be fair, early female drivers of gas-powered automobiles had major concerns that had to be overcome as well. One of those concerns was the availability of safe and attractive places to refuel. I am old enough that I remember gas stations being wallpapered with pictures of pinup girls with stacks of *Playboys* on the floor and a greasy-looking guy pointing to some unclean and oddly damp toilet in the back of the shop. It was creepy.

To make gas stations more family friendly, owners and nationwide chains started to both provide and promote their clean restrooms and other amenities, such as free coffee. Many credit the creation of clean restrooms at gas stations across the country as being one of the things that made driving long distances more comfortable for White women, since this amenity was often off limits to drivers of color in certain parts of the country.[20]

Today, along with women in sundresses, manufacturers are working to make EVs appeal to women. EVs, after all, are our best, most advanced, technology to combat climate change. Ford brilliantly put engineer Lisa Zhang in charge of the creation and rollout of the Ford F-150 Lightning Electric Truck. Zhang is a second-generation Ford employee who holds two master's degrees. In videos made by car influencers, she competently and impressively explains a vehicle that is in every way a workhorse, including scales to weigh payloads. She does so in a way that conforms to traditional gender norms, telling the *Detroit Free Press*, for example, that her most cherished moments came with taking care of her children, not by launching the most important Ford vehicle since the Model T.[21]

Almost pitch-perfect, Zhang only occasionally hits a flat note, such as when she tries to envision a woman sitting in this enormous vehicle, finishing emails on her laptop at school pickup, or suggests that the front trunk, known as a "frunk," is a perfect place for transporting flowers.

Only 16 percent of trucks are purchased by women, and there is little evidence that women are lining up for electric trucks, even though the screen at the front of the vehicle includes a few nifty games and a sketch pad to play with while waiting in school pickup lanes.

Remarkably, however, few if any EV makers are pushing hard on a pressure point that is close to women's, and especially mothers', view of the world. That is, the lack of heavily polluted exhaust that has been linked to birth defects, asthma, and learning deficits in children. Not to mention the effects of climate instability that will affect generations to come.

That may have to do with the decades-long struggle that the auto industry has put into denying that those effects were ever real. Car companies are in a bind. Admit that their product has been noxious and toxic and proclaim that they've invented something better. Or sidestep that in a dance about vehicle efficiency and admittedly cool features such as built-in refrigerators and generators powerful enough to light up the backyard in the case of a power outage. Not to mention those girls in sundresses, who look like they are off to have a fine time in their electric cars. Clothes and cars, after all, are a tried-and-true combo for manufacturers hoping to lure female drivers.

With waits as long as three years for some models, and new brands such as the Afeela set to debut in 2025, EVs are the new "it" car of the decade.[22] And, to be sure, the end of the loud internal combustion engine with its dangerous exhaust is something to cheer about.

Still, doubters of the technology make several convincing arguments, the most powerful of which is that a battery charged at home or at a charging station is likely just a different way of consuming fossil fuels. That is, the source of the power in most homes comes from some sort of fossil fuel generation, a coal fired power plant perhaps, meaning that electric cars are just consuming fossil fuels in a novel way. To be sure, there are now efforts to create vast solar farms to power electric vehicles—including one in my home state that will put solar power panels on the roofs of garages at major transportation hubs—an excellent start to collecting on the promise of the EV.

Critics also argue convincingly that the aging power grid most likely cannot sustain a swift move to electric vehicles, that mining practices used in some types of batteries are abhorrent, and that recycling facilities for the batteries, which can weigh several hundred pounds, are nonexistent, and therefore these new and improved vehicles represent just a different kind of environmental problem that will rear its ugly head, in about a decade or so, as those gigantic batteries go dead.

All of which is more or less true, and yet to raise these concerns is to be accused of being a technophobe or, in the words of a Twitter mob, spreading FUD: fear, uncertainty, and doubt. Still, even a true car lover like me has to ask: Is the EV just another car story we are being sold? Or would a hydrogen option, as some have suggested, be better? And, of course, why not public transportation?

Writing in his superb book *Auto Mania: Cars, Consumers, and the Environment*, author Tom McCarthy argues that "in the United States many people share the powerful cultural preference—a stubborn hope—that problems will solve themselves as individuals make free choices in the service of personal desires."[23] That's essentially the embedded hope with the EV: that climate change can be solved with better individual purchases and we can avoid public policy changes and investments in things like public transportation.

There is a lot riding on the electric vehicle, and a lot will have to happen in a short amount of time for the effort to be successful. Besides an enormous mining effort to search for lithium, gas stations will have to close, leaving those without the funds for a new car with fewer and fewer options and hundreds of sites where the stations once stood left abandoned, many with old, leaking underground tanks, a significant source of groundwater pollution in their own right. Moreover, the gasoline tax currently is a hefty source of funds for the maintenance of state and federal highways. How exactly the roads will be funded without gasoline sales is unclear. A mystery even.

To be fair, this tension is part of all technological innovation. Early adopters must go out on a limb and purchase an unproven technology. Then enough people must follow to allow for more investment in this technology and for the price of the innovation to drop substantially. Only then will the product improve and allow the vast majority of purchasers to be able to afford the new technology. That's how widespread technological change happens.

To get the ball rolling, women will have to choose to give up cars that work in order to opt for cars that they hope are better in some way for the environment but also work. In my own case, outfitting my home for an EV would require drilling through the foundation of my one-hundred-year-old house to install a charger. That kind of rewiring would add several thousand dollars to an already expensive vehicle purchase. Even as a woman with a deep concern for the environment, I can't easily see a way to buy and keep an electric car in my old home.

As Leo Marx, in *The Machine in The Garden: Technology and the Pastoral Ideal in America*, admonished us: "Man wants his motor-car, and enjoys it, but he believes that it is the spontaneous fruit of the Edenic tree."[24] Or, to put it another way, we've always believed that we have the right to harvest anything around us for our own pleasure. In short, we are unwilling or unable to come to grips with the fact that the automobile is a great user of materials and spoiler of the earth, no matter what kind of engine it has.

In this way, detractors say that the EV is simply another way of refusing to take stock of the damage that cars have done to the environment. Electrics, they argue, are just a dressed-up way of making the damage look acceptable. There is, after all, an electric version of the monstrously sized nine-thousand-pound Hummer in which the battery alone weighs almost three thousand pounds, according to Edmunds.[25] Is that a car aimed at making better environmental choices? Or is that simply someone hoping for a pass on their questionable choice of a vehicle?

"'People use fashion and taste to rehabilitate themselves or empower a larger project,' said Sophia Rosenfeld, professor of history at the University of Pennsylvania and the author of 'Democracy and Truth: A Short History.'" They use new technologies "'to whitewash themselves or a national culture or a set of business practices.'"[26]

It may be that the feeding frenzy over electrics is just that: a papering-over of the real crimes of the automobile. There is precedence for this. Early plastics were made because environmentalists at the time were rightly concerned about the use of animal parts such as elephant tusks and turtle shells for consumer items such as billiard balls and hair combs.

And we all know how poorly the story of modern-day plastics turns out.

As Albert Einstein is widely reported to have remarked, "Anyone who thinks science is trying to make human life easier or more pleasant is utterly mistaken."[27]

SIXTEEN

Tomorrow's Vehicle: Autonomous, Connected, Distracting, and Dangerous

The most idealist nations invent the most machines. America simply teems with mechanical inventions because nobody wants to do anything. They are idealists. Let a machine do the thing.
—D. H. Lawrence, *Studies in Classic American Literature*

Besides the changes to the way the car itself is powered, a whole new mobility ecosystem is emerging: a rapidly evolving and elaborately interconnected system of roads, vehicles, and sensors—the outlines of which are just beginning to reveal themselves. At this point in our technology-obsessed and auto-addicted society, it is almost impossible to imagine a future that does not include some kind of semiautonomous sensing vehicle operating in some new kind of way on roads especially modified for its existence. And it's equally impossible not to believe that women will pay for these vehicles with both their bodies and their wallets, just as we always have.

The first pedestrian ever killed by a car was a woman named Bridgett Driscoll. The mother of three was killed by a novice driver in a vehicle engineered for an 1896 exhibition in England. At the time, there were no more than a handful of cars on the road, and at the inquest into Driscoll's death, the coroner solemnly wished that no other person ever suffer her fate.

Fast forward to 2018, a year in which roughly thirty-five thousand Americans were killed in traffic accidents, and the year in which the first pedestrian death caused by a semiautonomous vehicle was recorded in Tempe, Arizona. Elaine Herzberg, forty-nine years old, was killed by an autonomous Uber test vehicle operated by Rafaela Vasquez late on a Sunday evening in March. Vasquez was driving a gray Volvo SUV equipped with cameras and sensors that allowed the vehicle to operate autonomously in certain conditions. Herzberg, who had drugs in her bloodstream at the time of her death and who had battled addiction on and off, was living at a campground near the accident site. She was walking her heavily laden bike across the road at a dangerous spot, a place so dangerous that the town had erected signs warning people homesteading in adjacent campgrounds not to cross there.

Uber had been aggressively racking up miles for months in Arizona in a rush to prove the viability and safety of its vehicles. According to *Wired*, the company had landed in Arizona after California revoked the registrations of their vehicles when the company refused to get testing permits. In the months before the crash, cost cutting at the company had reduced the number of employees in each car from a team of two to one solo operator. That meant that the Uber operator, Vasquez, was alone, and it was necessary for her to monitor both the systems' alerts from the car and messages from other operators while also scanning the road for dangers. In videos taken at the time of the crash, Vasquez was clearly distracted by something, and whether that was her own phone or the company's phone remains unclear.

What is clear, however, is that on that night, according to Uber's report to the National Transportation Safety Board, the self-driving system did not even realize Hertzberg was human. *Wired* reporter Lauren Smiley puts it this way:

> Nearly every time the system changed what it thought Herzberg was—a car, a bike, other—it started from scratch in calculating where the object might be headed, that is, across the road into the Volvo's Lane. Uber had programmed the car to delay hard braking for one second to allow the system to verify the emergency—and avoid false alarms—and for the human to take over. The system would brake hard only if it could entirely avoid the crash, otherwise it would slow down gradually and warn the operator. In other words, by the time it deemed it couldn't entirely avoid Hertzberg that night, the car didn't slam on the brakes, which might have made the impact less severe.[1]

Slamming on the brakes at the last minute would have been the responsibility of the operator, who on video appears to have been distracted by a device. According to some press reports, Vasquez was streaming *The Voice*, but because of the company's cost-cutting efforts she was also tasked with checking the company's chats on Slack. You can watch this for yourself, since remarkably, and somewhat unbelievably, we can watch this whole thing play out on the internet, complete with inside and outside views and the devastating second in which two innocent women's lives were forever upended by an unproven and wildly underregulated technology.

Uber reportedly settled with the victim's family within ten days, and it just as quickly distanced itself from Vasquez as she faced charges of negligent homicide with a dangerous instrument. Charges that might have resulted in four to eight years in prison if Vasquez had been found guilty. None of the possibly responsible executives or corporate entities, including Uber, the manufacturers of the sensors used by Uber, or the manufacturer of the vehicle itself, were charged in the crime, nor were they held publicly accountable. The state of Arizona, after luring Uber to its low regulatory environment, politely asked them to leave. The company went public the following year with a $75 billion valuation.

Five years after the crash, Vasquez agreed to a plea bargain that allowed her to avoid jail time. Unfortunately, the plea allowed all corporate parties and the nascent self-driving industry to avoid a trial that might have aired the many dangers of the new technology publicly.[2] It remains to be seen whether the case will set a precedent, one that holds "operators" responsible while letting technology makers off the hook.

Herzberg's death was an important inflection point in the rush to develop an autonomous car, and one that should give us all pause. We know for certain that the arrival of the Model T changed women's lives forever. Yet what is less clear is how the emerging sector of connected and semiautonomous vehicles will affect women. More importantly, perhaps, it is unclear how much impact and input women consumers will have in the way the technology is developed, even as it impacts our landscapes and our lives.

A woman sits atop General Motors and, as of this writing, women hold the top marketing jobs at all three of the major US manufacturers. Still, there continues to be plenty of evidence that a woman's point of view is not widely heard in the design studio or the boardroom, especially in the wild west of automotive startups. Consider that the EV startup Rivian managed to have a very public sex discrimination dustup even before it went public. Laura Schwab, who had decades of experience at car companies such as Aston Martin, says she was forced out after she called out the "bro culture" of the automaker.[3]

"Exclusion became a pattern, and I was left out of countless meetings where business needs and my role dictate that I should have been present," wrote Schwab. "I thought that my years of experience and my deep knowledge and expertise had earned me a spot at the table, but at Rivian it did not."[4]

Buyer, driver, operator, and vulnerable pedestrians everywhere, beware. Here comes a new era. One in which women will have more say than ever but still not enough to keep us safe.

❋

As of 2021, 90 percent of cars sold in the United States—and around 130 million cars sold worldwide—contained some form of embedded connectivity.[5] Moreover, the cars of tomorrow will be more like iPhones than they are like the cars my father sold. That is, they will be highly connected and technologically advanced. No longer will mechanics listen for the rattle or whine of the engine—now they will run proprietary computerized diagnostic tests from the cloud, and that means that increasingly you will not simply drop your car off at the corner garage to get fixed; it must go to the manufacturer, through either a dealer or a service center. In your next car when you want more horsepower or a more pleasing engine sound you will simply download it, like an app but with an enormous price tag. Ditto for any recalls that require changes in major systems. And when your car does need service, it will tell you and then help you book that service through the touchscreen on the console. Moreover, Tesla regularly updates its vehicles remotely, saving owners a trip to a service center.

But beyond basic functionality there will soon exist a whole new system of car electronics, interfaces, and economic partnerships that will use your car as home base. In fact, a new communications channel is already being prepared for a world in which your car senses the road, gauges the size of the parking spot at the grocery store, and records and remembers your preferred route to work.[6]

To facilitate this connectivity, in November 2020 the Federal Communications Commission voted to give part of the 5G spectrum to a different standard called "cellular vehicle to everything," or C-V2X.[7]

Of course, all this might mean better performance and ease of driving, but it comes with a cost. And this is where consumers—especially women—need to be particularly aware. As Victoria Scott says in an article for the website Drive, "While most owner concerns (and popular attention) have been fixed on unallowed hacks into such systems by bad actors, there are still massive troves of automatically generated data open to anyone with the knowledge to access it, and even the 'proper' use of this data can be a risk to consumers who seek privacy. Your home, your work, every trip you've taken no matter how private: it all can be seen by companies, countries, and individuals you've never given permission to follow your travels, and completely legally."[8]

This third party—be it someone well-known such as Google or a yet-unknown competitor—will know your habits: your visits to the grocery store and the bar. It will know which day care you drop off at and which shoe stores you've visited. Big Brother will be watching and collecting data at every stop, selling and reselling it over and over again, probably with your permission, as you most likely didn't scroll through the twenty pages of documentation before clicking ACCEPT.

"Google and Amazon are already locked in competition for the dashboard of your car," writes Shoshana Zuboff, a Harvard Business School professor who has coined a new term for the ways in which companies can capture data and trade that data in a new kind of marketplace. She calls it *surveillance capitalism.*

"Surveillance capitalism," she explains, "unilaterally claims private human experience as its own commodity that can be translated into behavioral data which can be then sold and purchased in a new kind of marketplace that trades exclusively in predictions of our future behavior, what we will do now, soon and later."[9]

For example, Facebook, or Meta as the company is now called, uses data such as which videos you watch to create targeted audiences for advertisers. "Information about a user's digital history—such as what videos on Instagram prompt a person to stop scrolling, or what types of links a person clicks when browsing their Facebook feeds—is used by marketers to get ads in front of people who are the most likely to buy," according to the *New York Times*. These practices helped Meta generate $118 billion in revenue in 2021.[10]

While we have all become accustomed to scrolling through an app on our phone regardless of the dangers that may be involved, things get a lot more complicated when your car is essentially one big app sending data to unknown parties about your whereabouts and habits.

This connectivity may be great when your car alerts the authorities that the bridge has washed out and you're about to drive over a cliff. But while this may have some advantages, the downsides could be enormous.

Behind on your car payments, as about 7 percent of people currently are?[11] Consider this scenario: Google's mastermind, Hal Varian, has observed that new vehicular monitoring systems will make it possible for car companies to repossess a car simply by remotely instructing it not to start and then instructing it to signal its location so it can be picked up.

Zuboff, a mother, reacts with appropriate horror to this idea. "What happens to the driver? What if there is a child in the car? Or a blizzard? Or a train to catch? Or a day-care center drop-off on the way to work? A mother on life support in the hospital miles away? A son waiting to be picked up at school."[12]

For most women, most of the time, cars are tools. We aren't headed out to joyride. In an era in which women's bodies are heavily politicized, does your car know that you've driven a friend—or your daughter or yourself—across state lines to an abortion clinic, a procedure that is not legal in the state in which your car is registered and tagged? Such a drive might make you an accomplice to an illegal procedure. So, who has access to that trip report? Who does your car alert about your trip to the abortion clinic?[13]

In another possible scenario, can your abusive spouse track your car? The car was one of the first routes of escape for abused women, but will that escape route still work if your husband can disable your car through an app on his phone? Then what? What will happen to women like my mother if they can be tracked through town? Will teenage kids ever know the whistling sound of freedom if their mother is using the devices on the family car to track their every move? So much has changed since the days when I could spin my car around and around in an empty parking lot after a light snow while the cops looked the other way and my boyfriend whooped from the passenger seat. Will kids ever know that special spark?

And who profits from all that data? Teslas and other high-end vehicles now have large touchscreens mounted to the dash between the seats. Ultimately, some of that space could be sold to advertisers. Is it possible you will have to scroll through six ads before your car will start? Yes. And for sure the baby will be screaming in the car seat, and your husband will be fuming trying to get to work. More stress for mother indeed.

Our attention is a commodity, and our time is precious. But those commodities are now being sold to the highest bidder without our ever knowing who is profiting. Nowhere is this clearer than in the development of the vehicle of tomorrow, and no one should care more about this than women, whose attention spans are already split between home and work and children and spouses and sick parents to a frazzling degree.

Matthew Crawford, author of *Why We Drive: A Philosophy of the Open Road*, is a University of Chicago–trained philosopher, auto mechanic, and motorcycle enthusiast. He longs for the open road, and he writes in alluring ways about the pleasures of leaning into a curve, of opening up the throttle and being in charge of both your route and your destiny.

It is no surprise that he sees the next generation of automobility as soulsucking. "By colonizing your commute—currently something you do, an actual activity in the tangible world that demands your attention—with yet another tether to the all-consuming logic of surveillance and profit, those precious fifty-two minutes of your attention are now available to be auctioned off to the highest bidder," he writes. [14]

"And the reason that you're spending thousands of dollars for sensors, and warning buzzers," says Jamie Kitman of *Car and Driver* magazine with some incredulity, "Is so you can be marketed at more?"

And you will likely buy into this. Why? Because it will be bundled with a dozen things you need. It will be put forth as necessary for safety, and before long it will be impossible to opt out of because it will literally be baked into your car. And, unlike your phone, you will not be able to turn off Wi-Fi because you will likely need it to access some sort of necessary or emergency feature.

❄

When I was a kid there was this cartoon set in 2062 called *The Jetsons*. The Jetsons were modern in every way even as they replicated the odd stereotypes of the past. Consider their car choices. They had a sort of low-flying airplane/car that seemed to drive itself. Enough so that they could all play a wholesome board game on their way to school. And you know what, Mrs. Jetson was actually a terrible driver.

Which is just to say that the fantasy of an automated vehicle has been around for a long time, just as long as the trope about women not being good drivers and knowing nothing about cars has. These prejudices never seem to die; they are biblical in the way they take over our imagination.

And in the future, it won't be some slick guy in the showroom selling us our cars. We channel all our choices now through sanitized clicks. Now you can book a virtual tour of the car you are interested in, order it online, and have it delivered to your home. The direct-to-consumer model is here to stay and can be profitable, as Tesla has shown. Other upstarts, such as Carvana, also make it possible to buy online in an orchestrated series of clicks.

On the plus side, this may mean less difficult showroom interactions for women. But the selling won't stop, it will simply move to the "curated content" created for social media. Buyer beware because that social influencer on Instagram is every bit as baked into the system as my father was.

Consider for a minute that Waymo, owned in part by Google, a leader in self-driving automation, recently published on their website a post titled "Mom Approved: Two Busy Parents Share Why They Think Autonomous Driving Could Help Families and Small Businesses."

The company smartly asked two Arizona moms to hop into Waymo's One, an autonomous ride-hailing service, from the Chandler Mall to the Porch restaurant. The two women, who call themselves the AZ Mom Squad, are part of a small army of women trying to eke out a living as influencers on social media by posting about cars and other business.

Women car influencers tend to be of a type. Say, an over-the-knee boot-wearing woman checking out the back of an Escalade to see if she can get the seats to fold down without breaking a nail. Or there is the mom-type influencer checking out the cupholders and the built-in refrigerators and pointing out something fun on the dashboard.

The AZ Mom Squad, Megan Ghormly and Ashlyn Leyba, seem like a nice pair of sisters who, according to their site, share a love for shopping at Target and can't go more than a day or two without getting together. On their trip in the Waymo automated car, the kids are well-behaved, and they say the ride is "shockingly cool." The car carefully navigates around another vehicle on the side of the road, and the kids are excited to be part of the adventure.

"My daughter could not wait to push the button that said, 'start ride,'" Megan says. "Then it started, and she screamed out the window she was so excited," Ashlyn recalls, adding that "one of the children nearly cried when the ride was over."[15]

As content creators, the women say, they are working constantly and don't have the time to drive everywhere. Megan emphasizes, "This would be so beneficial."

The last shoe has dropped. Women are now selling themselves the car of the future.

SEVENTEEN

Actual Miles May Vary: The Story We Tell Ourselves about the Car

The status of a whole civilization changes along with the way in which its everyday objects make themselves present and the way in which they are enjoyed.

—Jean Baudrillard, *The System of Objects*

In 2022, there were more cars than people in California. The Hummer EV had a waiting list of seventy-five thousand people, even though it will not be able to cross the Brooklyn Bridge as it is heavier than most commercial trucks.

During the midterm congressional elections, President Joe Biden hopped into a Ford truck to establish his credentials as a "car guy," and the minivan-driving mom briefly came back into focus as a swing group. The *New York Times* called her "a white, married, minivan-steering, cleats-toting, home-owning swing voter, exhausted by culture wars and seeking optimistic, common-sense politics."[1]

It sounded just about right and ever so much like the 1980s all over again. Especially the exhausted part.

Record heat waves, historic droughts, and enormous wildfires became potent reminders that gas-guzzling cars contribute to climate instability. The war in Ukraine drove gas prices to almost $5 a gallon in the summer of 2022, and all three major American car manufacturers committed to electric vehicles, even though just about everyone is concerned that the grid will be unable to handle it.

Post-pandemic many people continued to resist the notion of going back to the office, but inexplicably, in Boston at least, traffic seems to be worse than ever.

In short, nothing about the car makes sense—except that just about everyone needs one.

❊

Car myths are part of our American creation story; like Adam and Eve in the Garden of Eden, they tell us what it means to be male and female. But like a lot of myths, they've become distorted over time, used for commercial and political purposes to obscure and to profit.

I've tried to point out all the obvious and not-so-obvious ways we, as women, have accommodated the car. I've used my own personal experiences to make visible the many stories we have been told about the car. And I've taken a close look at the stories about the car of the future, about which I am both hopeful and skeptical.

The power of these stories resides not simply in their telling and retelling but also in the history they kick to the side of the road. That is, the things they highlight and the things they obscure. They keep us from seeing the carnage at the intersection and help us to forget the women whose lives have been lost to the cause. Bridget Driscoll, Elaine Herzberg, and Rafaela Vasquez, to name a few. The women who have died and will continue to die in the name of progress that may not really be progress at all, but simply a fantasy we keep selling ourselves about a better life through technology.

These stories tell us not just what to do with the technology, they also tell us how to feel about it. How the car can and should define us. But only rarely have women been the authors of their own stories about the car, especially in collective popular culture.

There is one notable exception. Consider the 1991 feminist road movie, *Thelma and Louise*, written by Callie Khouri. The plot is simple. Two women head out for a girl's weekend to escape their less-than-stellar lives and less-than-romantic lovers. After drinking in a bar, a man attempts to rape Thelma, violently pushing her up against the hood of his vehicle in a parking lot while she screams.

Louise, full of leftover rage from an undisclosed trauma, stumbles across the attack and pries Thelma away. But when the rapist lets loose a string of profanities, Louise shoots and kills him. Soon, the police are in hot pursuit. The women flee in their 1966 Ford Thunderbird, and the trip becomes a road movie extraordinaire, complete with an orgasmic hookup with the gorgeous young Brad Pitt and a comical scene where they lock a trooper in the trunk of his patrol car. Among their many adventures is a scene in which the women are harassed on the road by a man driving a fuel truck who repeatedly uses an obscene gesture to request a hand job.[2] Thelma and Louise confront their harasser on the side of the road. The man refuses to apologize, so they shoot out his tires. When he calls Louise a bitch, they blow up his truck, leaving him standing next to a blistering wreck.

That scene was cathartic for so many women of my generation who had had the experience of being leered at or harassed by men on the road. Few of us had the nerve to blow up their trucks, but many of us were thrilled when Thelma and Louise made their incendiary case for a little respect on the road.

At the end of the movie, Thelma and Louise, surrounded by police, choose to drive off a cliff rather than surrender to what we would now call the patriarchy, which has descended upon them in full force with flashing lights and guns drawn. No way were they going back to a world where men dominated and abused them on and off the road.

In *Why We Drive: Toward a Philosophy of the Open Road*, Matthew Crawford calls it an "inspired, terrible moment of self-destruction."[3] And it was. As they drove over that cliff, the women in the theater actually cheered. Some of us cried. Not because they died, but because they were active agents in their own lives. Feeling the power of the dainty toe on the accelerator, they were going to be in charge of their lives, even if that meant choosing to go over that cliff in a blaze of glory. They were behind the wheel. They were driving.

Now contrast this Thelma and Louise moment with the television show *American Auto*, which premiered on NBC in May 2022. *American Auto* is a workplace comedy that follows the employees of a fictional one-hundred-year-old company called Payne Motors. In the pilot episode, we're introduced to Katherine Hastings, the new CEO, who is not a Payne but is a woman.

Hastings knows nothing about cars and doesn't even know how to drive. The idea is to make her an awkward fish out of water, who is clueless in the way Michael Scott was clueless on *The Office*, or possibly in over her head like Selina Meyers on *Veep*.[4]

The company is about to launch the first self-driving car, and it's also Hastings's first day on the job. She and the team take a test run on a back lot, with Hastings in the driver's seat. She doesn't know how to operate the car—in this case command it to "go"—and when one team member refers to the car as *she*, there is a debate about whether or not they should be assigning a gender. This banter takes place as they drive but comes to a stop when they almost hit a cutout of a family crossing the street. The car safely stops and proceeds, only to hit a Payne employee a few seconds later. The gag is that the Payne employee hit was a Black person and the family the car did not hit was White. This joke highlights an actual problem with early technology and its inability to recognize darker skin tones.

All of this is baked into a television show and made to seem funny. Why do these stereotypes persist and why do people still think they are funny? In 2022 we are already setting the tones for a new era of automotive history, and the women are what, clueless? Mary Barra became CEO of General Motors in 2014. The person in charge of engineering of the Ford-150 is an Asian woman. Women have raced in the Indy 500 and are now a formidable force on the NASCAR circuit. Women drive every possible form of vehicle, from a tractor on the farm to a tractor trailer on the highway.

But to be sure, even high-minded cultural attempts at reinventing our thinking about cars can fall off course.

Consider the children's book *Cars! Cars! Cars!*, published by the Museum of Modern Art in connection with its *Automania* exhibit in 2021. In it we meet a young girl named Rosario. She loves cars! All her toys are cars, she draws cars, and she even has pictures of cars on her clothing. In the beginning of the book, she can't wait to drive. She even asks her parents if she can drive their car. No, she's like six or maybe eight years old, she can't drive the car.

But she retorts: I could do the grocery shopping and a list of other chores! Notice the default female position of making herself useful via her vehicle. *No, Rosario, no! Don't do it*.

So, then she thinks about a car race and thinks to herself how quickly she could do the grocery shopping. You've got to be kidding me. *Wait, really? Noooo*.

Finally, she figures out that she could make a great adventure in her car and longs for a Jeep. *Finally, Rosario! You go, girl!*

But wait, a smart car would be easier to park. But then she evokes her inner designer and decides to get the goddess car, the Citroën DS. Or maybe a Porsche or a ragtop Jaguar where she can let her hair roam free. *Now you're thinking, Rosario*. Then out comes her little hipster side, and she longs for an Airstream trailer that she can use to visit all the national parks. *Oh, the Instagram possibilities, Rosario.*

Then she quickly realizes that the car has changed our landscape forever. Not bad for an eight-year-old. But now she knows how much gas cars use and briefly longs for a Volkswagen Beetle. Good thing she doesn't know about its Nazi past.

So quickly she moves on to designing a car that is battery and solar powered. *Go, Rosaria.* But, wait. Everyone would want one of those, and how would that shake out? By the end of the book, she thinks cars are intrinsically bad—the smell and the exhaust. So, in conclusion: Rosaria does not get a car. She decides to take an all-electric bus, because it would fit so many more groceries.

Maybe this book, so carefully constructed by the staff at the Museum of Modern Art, challenged some of the ideas that we have about women and cars. Hey, at least she sort of fantasizes about having some fun. But in the end, it just reinforced so many more old and negative ideas that we have about women and cars. Why is an eight-year-old girl fantasizing about better ways to use vehicles to do the grocery shopping?

Sometimes, the more we fight against stereotypes, the more we reinforce them.

To be clear, the Museum of Modern Art has an amazing collection of cars, most of which were designed to be, well, gorgeous and aesthetically pleasing in so many ways, from line to form to color. They are exquisite and about so much more than getting the groceries. They are also chameleons with many different uses, shapes, and meanings. They are artwork.

Moreover, the car lies at the center of a complex ecosystem in which small changes anywhere can lead to big, unforeseen, almost inexplicable changes later on. Take those weird little plastic stoppers on Starbucks coffee cups—those are a direct descendant of the spills that burned a woman's lap, scalding her labia, in 1992. A small problem with the design in one vehicle added millions of pieces of small plastic to the world years later. The car is like that.

And sometimes car culture can just be inexplicable.

Eating ice cream in a small park in my town, I noticed a truck fly by with enormous gold truck nuts. I pointed and shouted to my husband the obvious: "Look, truck nuts!"

In case you don't know, truck nuts are just what you think they are. Plastic or metal appendages in the shape of a man's testicles that are attached to his truck.

And aren't truck nuts redundant if the truck itself is a kind of male appendage, the bigger the better? Sometimes you just can't make this stuff up. Often it is impossible to explain to anyone who wasn't born to it like I was.

For example, that same summer on another hot day, I slink into the air-conditioned post office to mail a package. Up on the wall I see a poster of new stamps, and what does my little eye spy but some of the cars of my youth.

Most notably the 1969 Ford Mustang Boss 302, 1970 Dodge Challenger R/T, 1969 Chevrolet Camaro Z/28, 1967 Mercury Cougar XR-7 GT, and 1969 AMC Javelin SST. The stamps are being marketed as Pony Car Stamps, which they are, in fact, except it's not a term most people use much anymore. [5]

I ask the woman at the counter for the Pony Car stamps. She is a nice professional woman with an accent—an immigrant. I've seen her before, and we have an easy rapport. She mistakenly brings me stamps with horses on them—not cars.

"No!" I shout through my mask and above the noise of the old air conditioner, "I want the Pony Cars."

"Huh?" she says. Now I am gesticulating wildly toward the poster, and we both start to laugh.

"Why do they call them Pony Cars?" she asks reasonably. "Are they named after horses?"

"Well, at least one of them is, sort of," I explain.

"Okay, okay," she says. "America, pretty crazy."

"Yes, pretty crazy," I say. "Car crazy!"

I buy five sheets of stamps to send to my car friends and thank her for the laugh.

Outside on the street, the cars whiz by way above the twenty-five mph speed limit that my town has declared but does not enforce. A recent traffic study showed thirty-two thousand cars passing by on a street adjacent to my home in a one-week period with a small percentage traveling well over the stated speed of twenty-five miles per hour.

Do you know how speed limits are set? It has nothing to do with the road or the number of cars. No, speed limits are set based on a metric. That is, traffic engineers work together to come up with a number that they believe 85 percent of the people will obey. Once they are convinced of that metric, they set the speed limit there.

Let's be clear that increased speeds harm pedestrians. According to engineer Charles L. Marohn, "For people outside a vehicle, the average risk of severe injury jumps from 50 percent at just 31 miles per hour to 75 percent at 49 miles per hour. The average risk of death for a person outside a vehicle jumps from 10 percent at impact speeds of 23 miles per hour, to 25 percent at 32 miles per hour and 50 percent at 42 miles per hour."[6]

I wrote this book in my tiny corner office, struggling with the noise from those passing cars. Noise from traffic has been linked with both higher blood pressure and with increased risk of Alzheimer's, probably due to disturbed sleep. Living near traffic is part of the cost of poverty for some.

My town is a small suburb close to a major metropolitan area. I've lived there for almost thirty years. When we first moved in there was very little traffic through the town. Then the GPS happened and now cars routinely cut down my street, which is only one block in length, guided by Waze or Apple.

At a recent town meeting held on Zoom, residents complained of the noise and the turmoil while town administrators struggled. They can only reroute traffic in ways that will make it worse for our neighbors, not better for everyone. We are in every way fighting against ourselves in a conflict that is being worked out somewhere on Google Maps or on Waze by an algorithm we do not understand and cannot access. It is every bit what it means to be modern.

As Crawford says in *Why We Drive*, "It's one thing to lose a fight, another to be imperceptibly suffocated under the insistence that there is no fight, only the operation of disinterested reason." Hannah Arendt, he says, "called this 'the rule of nobody.'" [7]

✻

In a bid for nostalgia and to boost town spirit, once a year my hometown of Waukegan, Illinois, puts on a small event for car aficionados. That is, they invite all the old greasers and hot-rod heads back to Scoop the Loop and remember the so-called good old days. The following day is a kind of car show where everyone lines up their rides and gets to show them off to people walking the long length of downtown.

Waukegan is not what it was when I was growing up and my dad was selling cars. The factories around the lake closed, and the Whites moved out, either to the emerging Sun Belt or north to Milwaukee. That's when the Mexicans moved in. "Too many to count," my dad used to say in his unkind way. Today my hometown is mostly Mexican.

The great thing about the new Scoop the Loop is that it is a celebration that crosses both cultures. There are the Mexicans' cars with bright paint jobs, maybe a fur interior and some sort of statue on the dash. Then there are the guys I went to high school with; some of them have big beer bellies now and many are lacking hair, but their cars are still as shiny clean and stunning as they were in high school.

It's a kind of small miracle that for one day everyone comes together in support of the town's past and present and the cars that bind us. When I attended the event in 2022, I got to talking with a couple of women there. The first woman was my age, and she had just lost her husband. She had his old Mustang completely refurbished to honor him, and in a unique twist had his ashes mixed into the paint, which was the same Michigan blue as his alma mater. It was stunning and she won a trophy for her work on it.

Another was a thirty-year-old Mexican American woman with two degrees from Tufts University, dark loose curls, and a winning smile. Dulce and her father are members of a car club called Latin Cruisers. Her father bought the car in 1995 just after she was born. It was a 1969 blue Chevy Impala reminiscent of my mother's.

What did the car mean to her? I asked. She said it was a testament to their new home, their belonging in a new culture, the fact that they were creating and remaking the town even as they were joining it.

She splits the costs of the car with her dad. They were thinking of redoing the interior. I encouraged her not to. It's perfect, I said, just like my mom's.

If my childhood was spent keeping track of my father's lies, my adult life was spent trying to make sense of the world. Story by story, fact by fact, and sometimes comma by comma. Trying to understand how the complex global auto industry shapes our society even as our own choices and purchases shape the industry. Working to define how the car and the automobile culture of my youth—such a central part of my life—shaped me, beginning with my father's infectious love for it, my mother's sense of romance through it, my own unmistakable feeling of power and freedom every time I got behind the wheel. And the ever-growing sense—as I became an environmental reporter, a mother, a cancer survivor, and then the mother of a cancer survivor—that the heavily romanticized car was ruining just about everything.

In my work as a reporter, I began to see an orchestrated symphony of falsification, deception, and obfuscation at all levels of business and government. There were the major chords of corporate scandal and the minor notes of misinformation we now know as marketing. And when is marketing any slyer than when it involves women and cars? Consider the second *Sex in the City* movie, released in 2010. Samantha achieves an orgasm atop her Mercedes G Series as part of a promotion for the brand. Samantha is seen having loud, enthusiastic sex on the Fourth of July on the hood of the Mercedes while Carrie says via voiceover, "In the land of the free, and the home of the hormones," to coincide with the camera panning to fireworks as we hear Samantha climax. Or, as they say, he hit her G spot on top of her G Wagon. An acrobatic feat if ever there was one.

For the promotion of the film, Mercedes was hoping to capitalize on the trendsetting nature of the show; people bought the shoes, the clothes, the cupcakes. One mention could sell out a restaurant. Why not cars? "You can sell anything if you're shameless enough," a marketing professor from Harvard told me once.

"I know; my father was a used car salesman," I said.

He laughed. He thought I was kidding.

❇

At the end of the day, as Scoop the Loop was coming to a close, I said goodbye to Dulce and waved to a few friends from high school. I headed to an inexplicable mile of highway laid between two exits that literally goes nowhere. Today, it is often used for movies. Back in the day, we would drag race there.

I opened up my little rental car, but its lack of horsepower was a disappointment. No matter. I crossed an old bridge and remembered an old boyfriend and a stolen kiss at dawn.

I raced past the old factories and remembered the smell of the smoke from the tannery and drove past the now closed coal-fired power plant whose decades-old pollutants seep slowly toward the lake.

I thought about all that the car has done for us and done to us.

As one writer said about the effects of the car, once you see them, you can't not see them.

And you likely can't escape them either. No matter how fast you drive.

Acknowledgments

Books of this kind cannot be written without the assistance of many institutions and individuals.

I received help from many different sources including residencies at Ragdale, The Mesa Refuge, The Blue Mountain Center, and The Vermont Studio Center.

Researchers at the New York Public Library, the Detroit Public Library, the GM Heritage Center, the Henry Ford Museum of American Innovation, and the Lake County Historical Society all gave generously of their time.

My agent, Reiko Davis at DeFiore & Company, saw the promise of this book and helped shape it. My editor, Jessica Case, bravely took the book on. Alexis Rizzuto worked on several early drafts. Copyeditor J. LeTourneur Bax helped avoid many missteps. Rochelle Roberts provided much-needed administrative and research support. Nicole Maher of Pegasus worked tirelessly to get attention for this book.

An early version of this work was edited by Hattie Fletcher and published by *True Story*, a publication of *Creative Nonfiction Magazine*.

Finally, I would like to thank my husband, Jon, for his support, love, and patience.

Bibliography

"2020 Fatality Data Show Increased Traffic Fatalities during Pandemic." NHTSA, June 3, 2012. https://www.nhtsa.gov/press-releases/2020-fatality -data-show-increased-traffic-fatalities-during-pandemic.

Adkins, Jan. *Bertha Takes a Drive: How the Benz Automobile Changed the World.* Watertown, MA: Charlesbridge, 2017.

Akhim, Eugenia. "The Ugly Truth behind the Fastest Cannonball Run Ever." HotCars, June 10, 2022. https://www.hotcars.com/fastest-cannonball-run/.

al-Sharif, Manal. *Daring to Drive: A Saudi Woman's Awakening.* New York: Simon & Schuster, 2017.

Albert, Daniel M. *Are We There Yet? The American Automobile, Past, Present, and Driverless.* New York: W. W. Norton, 2019.

"Americans Are Borrowing at Record Levels to Pay for Their Expensive Cars." CNN Business, August 25, 2022. https://www.cnn.com/2022/08/25/cars /car-price-borrowing/index.html.

Appleyard, Bryan. *The Car: The Rise and Fall of the Machine that Made the Modern World.* New York: Pegasus Books, 2022.

Automania. MoMA. https://www.moma.org/calendar/exhibitions/5210.

Automobile Garments and Requisites: Imported and Domestic Models for Men and Women. New York: Saks and Company, 1904.

"*Baboon and Young.*" Wikipedia. https://en.wikipedia.org/wiki/Baboon _and_Young.

Backman, Maurie. "New Data Reveals That Car Ownership Costs Women Up To $7,800 More than Men." Motley Fool, March 26, 2022. https://www

.nasdaq.com/articles/new-data-reveals-that-car-ownership-costs-women
-up-to-%247800-more-than-men.

Badger, Emily, and Alicia Parlapiano. "The Exceptionally American Problem of
Rising Roadway Deaths." *New York Times*, November 27, 2022. https
://www.nytimes.com/2022/11/27/upshot/road-deaths-pedestrians-cyclists
.html#:~:text=In%202021%2C%20nearly%2043%2C000%20
people,—%20cyclists%2C%20motorcyclists%2C%20pedestrians.

Bailey, Beth L. *From Front Porch to Back Seat: Courtship in Twentieth-Century
America*. Baltimore: Johns Hopkins Press, 1989.

Ballard, J. G. *Crash*. New York: Noonday Press, 1996.

Banham, Reyner. *Los Angeles: The Architecture of Four Ecologies*. University of
California Press, 2009.

Barry, Keith. "The Crash Test Bias: How Male-Focused Testing Puts Female
Drivers at Risk." *Consumer Reports*, October 23, 2019. https://www
.consumerreports.org/car-safety/crash-test-bias-how-male-focused
-testing-puts-female-drivers-at-risk/.

Barthes, Roland. "A Cruel Country." *New Yorker*, September 6, 2010. https
://www.newyorker.com/magazine/2010/09/13/a-cruel-country.

———. *Mythologies*. New York: Hill & Wang, 1957.

Barton, Bruce. *The Man Nobody Knows: A Discovery of Jesus*. Indianapolis:
Bobbs-Merrill, 1925.

"Batteries for Electric Vehicles." US Department of Energy Alternative Fuels
Data Center. https://afdc.energy.gov/vehicles/electric_batteries.html.

Baudrillard, Jean. *The System of Objects*. London: Verso, 2005.

Baxter, Charles. *Burning Down the House: Essays on Fiction*. Minneapolis:
Graywolf Press, 1998.

Bayley, Stephen. *Age of Combustion*. London: Circa Press, 2021.

Beck, Julie. "The Decline of the Driver's License." *Atlantic*, January 22, 2016.
https://www.theatlantic.com/technology/archive/2016/01/the-decline-of
-the-drivers-license/425169/.

Beck, Martha N. *Breaking Point: Why Women Fall Apart and How They Can
Re-Create Their Lives*. New York: Crown, 1997.

Behling, Laura L. "'The Woman at the Wheel': Marketing Ideal Womanhood,
1915–1934." *Journal of American Culture* 20, no. 3 (Fall 1997): 13–30.

Belasco, Warren James. *Americans on the Road: From Autocamp to Motel, 1910–
1945*. Cambridge, MA: MIT Press, 1979.

Bell, Genevieve. "Unpacking Cars: Doing Anthro at Intel." *AnthroNotes* 32
(Fall 2011): 1–6.

Berger, Michael. "Women Drivers!: The Emergence of Folklore and Stereotypic Opinions Concerning Feminine Automotive Behavior." *Women's Studies International Forum* 9, no. 3 (1986): 257–63.

Bouzanquet, Jean-François. *Fast Ladies: Female Racing Drivers 1888–1970.* Poundbury, UK: Veloce Publishing, 2009.

Bradbury, Ray. *Bradbury Stories: 100 of His Most Celebrated Tales.* New York: HarperCollins, 2005.

Bradsher, Keith. *High and Mighty: The Dangerous Rise of the SUV.* New York: PublicAffairs, 2003.

Brinker, Andrew. "'Really Concerning Behavior': Continuing Pandemic Trend, Traffic Deaths Rose Again in 2021." *Boston Globe*, January 23, 2022. https://www.bostonglobe.com/2022/01/23/metro/really-concerning -behavior-continuing-pandemic-trend-traffic-deaths-rose-again-2021 /?p1=BGSearch_Overlay_Results.

Brottman, Mikita, ed. *Car Crash Culture.* London: Palgrave Macmillan, 2002.

Bubola, Emma. "With Sensors on Streets, France Takes Aim at 'Noise from Hell.'" *New York Times*, February 21, 2022. https://www.nytimes.com /2022/02/21/world/europe/france-street-noise.html.

Butler, Judith. *Gender Trouble: Feminism and the Subversion of Identity.* London: Routledge, 2006.

"Buy Now, Pay Later: A History of Personal Credit." Harvard Business School Historical Collections. https://www.library.hbs.edu/hc/credit /credit4d.html.

Caitlyn Jenner's YouTube channel. https://www.youtube.com/watch?v=q -vJ1xtksEM.

Calder, Lendol. *Financing the American Dream: A Cultural History of Consumer Credit.* Princeton, NJ: Princeton University Press, 1999.

"California Moves to Accelerate to 100% New Zero-Emission Vehicle Sales by 2035." California Air Resources Board, August 25, 2022. https://ww2.arb .ca.gov/news/california-moves-accelerate-100-new-zero-emission-vehicle -sales-2035.

Carroll, Thomas. "Women, Cars and Liberation." *Monmouth*, Fall 2017. https ://www.monmouth.edu/magazine/women-cars-liberation/.

Clarke, Deborah. *Driving Women: Fiction and Automobile Culture in Twentieth-Century America.* Baltimore: Johns Hopkins University Press, 2007.

Clarsen, Georgine. *Eat My Dust: Early Women Motorists.* Baltimore: Johns Hopkins University Press, 2008.

———. "The 'Dainty Female Toe' and the 'Brawny Male Arm': Conceptions of Bodies and Power in Automobile Technology." *Australian Feminist Studies* 15, no. 32 (2000): 153–63.

Colias, Mike, and Nora Eckert. "A New Brand of Sticker Shock Hits the Car Market." *Wall Street Journal*, February 26, 2022. https://www.wsj.com /articles/a-new-brand-of-sticker-shock-hits-the-car-market-11645851648 ?mod=hp_lead_pos11.

"Columbia Electric Key Facts." National Motor Museum. https://national motormuseum.org.uk/vehicle-collection/columbia-electric/.

Cowan, Ruth Schwartz. *More Work for Mother: The Ironies of Household Technology from the Open Hearth to the Microwave*. New York: Basic Books, 1983.

Cowper-Coles, Sherard. "Secrets of a Diplomat." *Times*, October 14, 2012. https://www.thetimes.co.uk/article/secrets-of-a-diplomat-7kcxbjkksqq.

Crawford, Matthew B. *Shop Class as Soulcraft: An Inquiry into the Value of Work*. New York: Penguin Books, 2009.

———. *Why We Drive: Toward a Philosophy of the Open Road*. Boston: Mariner Books, 2020.

———. *The World beyond Your Head: On Becoming an Individual in an Age of Distraction*. New York: Farrar, Straus and Giroux, 2015.

Criado Perez, Caroline. *Invisible Women: Data Bias in a World Designed for Men*. New York: Abrams, 2019.

Curtiss, Aaron. "The Freeway Numbers Game." *Los Angeles Times*, April 19, 1996. https://www.latimes.com/archives/la-xpm-1996-04-19-me-60325 -story.html.

"Dashcam Video from inside Driver's Car Shows Him Shooting during Road Rage Incident." YouTube, January 28, 2022. https://www.youtube.com /watch?v=saQ72NZtrS0.

Davis, River, and Sean McLain. "Toyota Chief Says 'Silent Majority' Has Doubts about Pursuing Only EVs." *Wall Street Journal*, December 18, 2022. https://www.wsj.com/articles/toyota-president-says-silent-majority -has-doubts-about-pursuing-only-evs-11671372223?mod=hp_lead_pos5.

Dawson, Chester. "Is It Possible to Have Too Many Cup Holders? Auto Makers Are Trying to Find Out." *Wall Street Journal*, February 15, 2018. https://www.wsj.com/articles/need-a-place-to-park-your-drink-the-worlds -auto-makers-are-on-it-1518710790.

Dean, Sam. "The History of the Car Cup Holder." *Bon Appétit*, February 18, 2013. https://www.bonappetit.com/trends/article/the-history-of-the-car -cup-holder.

Dettelbach, Cynthia G. *In the Driver's Seat: The Automobile in American Literature and Popular Culture*. Westport, CT: Greenwood Press, 1976.

Dichter, Ernest. *The Strategy of Desire*. London: Taylor & Francis, 2017.

Dickens, Charles. *A Tale of Two Cities*. New York: Penguin Books, 2007. https://www.google.com/books/edition/A_Tale_of_Two_Cities /j_5UbgB_9isC?hl=en&gbpv=1&bsq=ever%20in%20the%20way.

Dizik, Alina. "Parents Pile into Work Conferences to Escape Their Families." *Wall Street Journal*, July 20, 2022. https://www.wsj.com /articles/parents-pile-into-work-conferences-to-escape-their-families -11658323598?mod=hp_featst_pos5.

Dolle, Tom, and Jeff Stork. *Glamour Road: How Fashion and Styling Defined the Mid-Century Automobile*. Atglen, PA: Schiffer Publishing, 2022.

Downs, Anthony. *Still Stuck in Traffic: Coping with Peak-Hour Traffic Congestion*. Washington, DC: Brookings Institution Press, 2005.

"The DS by Roland Barthes." Citroënët. http://www.citroenet.org.uk /passenger-cars/michelin/ds/32.html.

Dubner, Stephen J. "Are Women Being Taken for a Ride?" Freakonomics, January 11, 2010. https://freakonomics.com/2010/01/are-women-being -taken-for-a-ride/.

Eastman, Joel W. *Styling vs. Safety: The American Automobile Industry and the Development of Automotive Safety, 1900–1966*. Lanham, MD: University Press of America, 1984.

"Electric Vehicles." IEA. https://www.iea.org/reports/electric-vehicles.

Elkin, Lauren. *Flâneuse: Women Walk the City in Paris, New York, Tokyo, Venice, and London*. New York: Farrar, Straus and Giroux, 2017.

Ewing, Jack, and Peter Eavis. "Electric Vehicles Start to Enter the Car-Buying Mainstream." *New York Times*, November 13, 2022. https ://www.nytimes.com/2022/11/13/business/electric-vehicles-buyers -mainstream.html.

Falconer, Tim. *Drive: A Road Trip through Our Complicated Affair with the Automobile*. Toronto: Penguin Canada, 2008.

Featherston, Emily. "Federal Agency Touts 'Accelerated Efforts' to Implement More-Advanced Crash Test Dummies, Promises Continued Research into Safety Disparities." *InVestigate*, October 21, 2022. https://www .investigatetv.com/2022/10/21/federal-agency-touts-accelerated-efforts -implement-more-advanced-crash-test-dummies-promises-continued -research-into-safety-disparities/.

"Female Crash Fatality Risk Relative to Males for Similar Physical Impacts."
    NHTSA, August 2022. https://crashstats.nhtsa.dot.gov/Api/Public
    /ViewPublication/813358.

Fitzgerald, F. Scott. *The Great Gatsby*. Wisehouse Classics Edition, 1925.

Flink, James J. *The Automobile Age*. Cambridge, MA: MIT Press, 1990.

———. *The Car Culture*. Cambridge, MA: MIT Press, 1987.

———. "The Three Stages of American Automobile Consciousness." *American
    Quarterly* 24, no. 4 (October 1972): 451–73.

Ford, Anne, and Charlotte Ford. *How to Love the Car in Your Life*. New York:
    Mobius Books, 1980.

"Ford Motor Credit Marks 50th Anniversary." *F & I and Showroom*, August 20,
    2009. https://www.fi-magazine.com/317887/ford-motor-credit-marks
    -50th-anniversary.

Forster, E. M. *The Machine Stops*. London: Penguin Books, 2011.

Foster, Mark S. *Nation on Wheels: The Automobile Culture in America since 1945*.
    Belmont, CA: Thomson Wadsworth, 2003.

Fria, Robert A. *Mustang Genesis: The Creation of the Pony Car*. Jefferson, NC:
    McFarland & Company, 2010.

Friedman, Vanessa. "When the Czarinas Ruled the Front Row." *New York
    Times*, July 4, 2022. https://www.nytimes.com/2022/07/04/style/mira
    -duma-elena-perminova-ukraine.html.

Friedman, Walter A. *Birth of a Salesman: The Transformation of Selling in
    America*. Cambridge, MA: Harvard University Press, 2004.

Gastelu, Gary. "Jeep Named Most Patriotic Automaker and Brand." Fox News,
    July 4, 2022, https://www.foxnews.com/auto/jeep-most-patriotic
    -automaker-brand.

Genat, Robert. *The American Car Dealership*. St. Paul, MN: MBI Publishing, 1999.

General Motors. *A Power Primer: An Introduction to the Internal Combustion
    Engine*. Detroit: General Motors, 1955.

Goldstein, Dana, Robert Gebeloff, Allison McCann, and Brent McDonald.
    "Suburban Women, No Longer 'Soccer Moms,' Hold Key to Midterms."
    New York Times, November 4, 2022. https://www
    .nytimes.com/2022/11/04/us/suburban-women-midterms.html.

Gopnik, Adam. "Jane Jacob's Street Smarts." *New Yorker*, September 26, 2016.
    https://www.newyorker.com/magazine/2016/09/26/jane-jacobs-street
    -smarts.

Gordon, Robert J. *The Rise and Fall of American Growth: The U.S. Standard of
    Living since the Civil War*. Princeton, NJ: Princeton University Press, 2016.

Grinys, Aurimas. "The Citroën DS: A Goddess Ahead of Its Time." Dyler, March 21, 2022. https://dyler.com/posts/105/the-citroen-ds-a-goddess -ahead-of-its-time.

Guerrieri, Vince. "Jordan Motor Car Co.'s Famous Ad." *Ohio Magazine*, February 2018. https://www.ohiomagazine.com/ohio-life/article/driving-inspiration.

Guillory, Daniel. "Bel Air: The Automobile as Art Object." *Michigan Quarterly Review* 19, no. 4 (Fall 1980): 657–66. https://quod.lib.umich.edu/m /mqrarchive/act2080.0019.004?xc=1&g=mqrg&node=act2080.0019.004 :31&view=image&seq=239&size=100.

Gurwell, Matt. "Meet Sierra Sam—The Original Crash Dummy." *Legal Examiner*, March 23, 2022. https://affiliates.legalexaminer.com /transportation/sierra-sam/.

Halliday, Aria S. *Buy Black: How Black Women Transformed US Pop Culture*. Urbana: University of Illinois Press, 2022.

Harper, Mary Walker. "The Woman Who Drives a Car: Should Know How to Buy and Run One." *Ladies' Home Journal*, September 1915.

Hazleton, Lesley. *Everything Women Always Wanted to Know about Cars: But Didn't Know Who to Ask*. 1st ed, New York: Doubleday, 1995.

Heitmann, John Alfred. *The Automobile and American Life*. Jefferson, NC: McFarland & Company, 2009.

"A Heritage of Heroes, 1940–1949." Jeep, n.d. https://www.jeep.com/history /1940s.html.

Heys, Ed. "Cadillacs Crown Jewels." *Hemmings*, September 23, 2018. https ://www.hemmings.com/stories/article/cadillacs-crown-jewels.

Hickey, Dave. *The Invisible Dragon: Essays on Beauty*. Revised and expanded ed. Chicago: University of Chicago Press, 2012.

Hiller, Jennifer. "Tesla to Open Some of Its Charging Network to Other EVs, White House Says." *Wall Street Journal*, February 15, 2023. https://www.wsj.com/articles/tesla-to-open-some-superchargers-to -other-vehicles-white-house-says-2ed35fd5?mod=hp_lista_pos4.

Hiott, Andrea. *Thinking Small: The Long, Strange Trip of the Volkswagen Beetle*. New York: Ballantine Books, 2012.

Hochschild, Arlie, and Anne Machung. *The Second Shift: Working Families and the Revolution at Home*. New York: Penguin Books, 2012.

"Honesty/Ethics in Professions." Gallup. https://news.gallup.com/poll/1654 /honesty-ethics-professions.aspx.

Howerton, Kristen. *Rage against the Minivan: Learning to Parent without Perfection*. New York: Penguin Random House, 2020.

Iacovetta, Franca, and Mariana Valverde, eds. *Gender Conflicts: New Essays in Women's History*. Toronto: University of Toronto Press, 1992.

Ingrassia, Paul, and Joseph B. White. *Comeback: The Fall and Rise of the American Automobile Industry*. New York: Simon & Schuster, 1995.

Jackson, Lauren Michele. "The Zora Neal Hurston We Don't Talk About." *New Yorker*, February 14, 2002. https://www.newyorker.com/books /page-turner/the-zora-neale-hurston-we-dont-talk-about.

Jacobs, Jane. The Death and Life of Great American Cities. London: Random House, 1961.

Kay, Jane Holtz. *Asphalt Nation: How the Automobile Took Over America, and How We Can Take It Back*. Berkeley: University of California Press, 1998.

Keats, John. *The Insolent Chariots*. Philadelphia: Lippincott, 1958.

Kellan, Kate. "Good Vibrations? COVID Quiet Time Soothes Earth's Seismic Shakes." Reuters, July 23, 2020. https://www.reuters.com/article /us-health-coronavirus-vibrations/good-vibrations-covid-quiet-time -soothes-earths-seismic-shakes-idUSKCN24O2P2.

Kinchin, Juliet. *Automania*. New York: Museum of Modern Art, 2021.

Kirsch, David A. *The Electric Vehicle and the Burden of History*. New Brunswick, NJ: Rutgers University Press, 2000.

Kitman, Jamie. "A Brief History of Gasoline: Searching for the Magic Bullet." *Jalopnik*, January 28, 2022. https://jalopnik.com/a-brief-history-of -gasoline-searching-for-the-magic-bu-1848438134.

Knoedelseder, William. *Fins: Harley Earl, the Rise of General Motors, and the Glory Days of Detroit*. New York: Harper Business, 2019.

Korosec, Kirsten. "Rivan Hit with Gender Discrimination Lawsuit That Alleges Toxic 'Bro Culture.'" *TechCrunch*, November 4, 2021. https ://techcrunch.com/2021/11/04/rivian-hit-with-gender-discrimination -lawsuit-that-alleges-toxic-bro-culture/.

Kuhn, Maria Weston. "It's Time to End Discrimination in Crash Testing." *Ms.*, August 10, 2022. https://msmagazine.com/2022/08/10 /discrimination-women-crash-test-dummies-car/.

Kurczewski, Nick. "The Science of the New-Car Smell." *Car and Driver*, July 10, 2021. https://www.caranddriver.com/features/a36970626 /science-new-car-smell/.

Kushner, Rachel. *The Hard Crowd: Essays 2000–2020*. New York: Scribner, 2021.

Ladd, Brian. *Autophobia: Love and Hate in the Automotive Age*. Chicago: University of Chicago Press, 2011.

"Lawmakers Push DOT to Update Female Crash Test Dummies."
    Safety+Health, May 2, 2022. https://www.safetyandhealthmagazine.com
    /articles/22541-lawmakers-push-dot-to-update-female-crash-test-dummies.

Lawrence, David Herbert. *Studies in Classic American Literature*. New York:
    T. Seltzer, 1923.

Lawson, Helene M. *Ladies on the Lot: Women, Car Sales, and the Pursuit of the
    American Dream*. Lanham, MD: Rowman & Littlefield, 2000.

Leonhardt, David. "Race, Class and Traffic Deaths." *New York Times*, August 23,
    2022. https://www.nytimes.com/2022/08/23/briefing/traffic-deaths
    -class-race-covid.html.

Levitt, Dorothy. *Woman and the Car: A Chatty Little Handbook for All Women
    Who Motor or Who Want to Motor*. London: John Lane, 1909.

Lewis, David L., and Laurence Goldstein, eds. *The Automobile and American
    Culture*. Ann Arbor: University of Michigan Press, 1983.

Lewis, Sinclair. *Babbitt*. New York: Bantam Dell, 1998.

Lewis, Tom. *Divided Highways: Building the Interstate Highways, Transforming
    American Life*. Revised ed. Ithaca, NY: Cornell University Press, 2013.

Lezotte, Chris. "The Evolution of the 'Chick Car' Or: What Came First, the
    Chick or the Car?" *Journal of Popular Culture* 45, no. 3 (2012): 516–31.
    https://onlinelibrary.wiley.com/doi/10.1111/j.1540-5931.2012.00942.x.

———. "Have You Heard the One about the Woman Driver? Chicks, Muscle,
    Pickups, and the Reimagining of the Woman behind the Wheel." PhD
    diss., Bowling Green State University, 2015.

———. "What Would Miss Daisy Drive? The Road Trip Film, the
    Automobile, and the Woman behind the Wheel." *Journal of Popular Culture*
    54, no. 5 (October 2021): 987–1011. https://doi.org/10.1111/jpcu.13061.

"Liebeck v. McDonald's Restaurants." Wikipedia. https://en.wikipedia.org/wiki
    /Liebeck_v._McDonald%27s_Restaurants.

"List of Vehicle-Ramming Incidents during George Floyd Protests."
    Wikipedia. https://en.wikipedia.org/wiki/List_of_vehicle-ramming
    _incidents_during_George_Floyd_protests.

Lucsko, David N. *Junkyards, Gearheads, and Rust: Salvaging the Automotive Past*.
    Baltimore: Johns Hopkins University Press, 2016.

Macy, Sue. *Motor Girls: How Women Took the Wheel and Drove Boldly into the
    Twentieth Century*. Washington, DC: National Geographic Partners,
    2017.

Mangla, Ravi. "The Secret History of Jaywalking: The Disturbing Reason It Was
    Outlawed—and Why We Should Lift the Ban." *Salon*, August 20, 2015.

https://www.salon.com/2015/08/20/the_secret_history_of_jaywalking_the
_disturbing_reason_it_was_outlawed_and_why_we_should_lift_the_ban/.

Marchand, Roland. *Advertising the American Dream: Making Way for Modernity, 1920–1940*. Berkeley: University of California Press, 1985.

Marci, Carl D. *Rewired: Protecting Your Brain in the Digital Age*. Cambridge, MA: Harvard University Press, 2022.

Marohn, Charles L. *Confessions of a Recovering Engineer: The Strong Towns Vision for Transportation in the Next American City*. Hoboken, NJ: Wiley, 2021.

Marsh, Peter, and Peter Collett. *Driving Passion: The Psychology of the Car*. London: Faber and Faber, 1987.

Marx, Leo. *The Machine in the Garden: Technology and the Pastoral Ideal in America*. New York: Oxford University Press, 1964. Reprint, 1979.

Maushart, Susan. *The Mask of Motherhood: How Becoming a Mother Changes Everything and Why We Pretend It Doesn't*. New York: Penguin Books, 2000.

Mayyasi, Alex. "How Subarus Came to Be Seen as Cars for Lesbians." *Atlantic*, June 22, 2016. https://www.theatlantic.com/business/archive/2016/06 /how-subarus-came-to-be-seen-as-cars-for-lesbians/488042/.

McCarthy, Todd. *Fast Women: The Legendary Ladies of Racing*. New York: Miramax Books, 2007.

McCarthy, Tom. *Auto Mania: Cars, Consumers, and the Environment*. New Haven, CT: Yale University Press, 2007.

———. "The Black Box in the Garden: Consumers and the Environment." In *A Companion to American Environmental History*, edited by Douglas Cazaux Sackman, 304–24. Hoboken, NJ: Wiley, 2010. https://doi.org /10.1002/9781444323610.ch16.

McConnell, Curt. *A Reliable Car and a Woman Who Knows It: The First Coast-to-Coast Auto Trips by Women, 1899–1916*. Jefferson, NC: McFarland & Company, 2000.

McGee, Kate. "911 Transcripts Filed in Updated 'Trump Train' Lawsuit Reveal San Marcos Police Refused to Send Escort to Biden Bus." *Texas Tribune*, October 29, 2021. https://www.texastribune.org/2021/10/29 /trump-train-texas-highway-crash-police/.

Meyersohn, Nathaniel. "Believe It or Not, Gas Stations Used to Be Squeaky Clean: Here's What Changed." CNN Business, June 4, 2022. https://www.cnn.com/2022/06/04/business/gas-station-bathrooms -history.

Miller, Caleb. "Dodge Charger Daytona EV Concept Back for SEMA, and an Even Wilder Banshee Is Coming." *Car and Driver*, October 31, 2022.

https://www.caranddriver.com/news/a41823742/dodge-charger-daytona
-ev-concept-sema-power-details/.

Milosz, Czeslaw. *Collected Poems*. New York: HarperCollins, 1988.

Mintz, Steven, and Susan Kellogg. *Domestic Revolutions: A Social History of American Family Life*. New York: Simon & Schuster, 1989.

Mitchell, William J., Chris E. Borroni-Bird, and Lawrence D. Burns. *Reinventing the Automobile: Personal Urban Mobility for the 21st Century*. Cambridge, MA: MIT Press, 2010.

Molinari, Susan, and Beth Brooke. "Women Are More Likely to Die or Be Injured in Car Crashes: There's a Simple Reason Why." *Washington Post*, December 12, 2021. https://www.washingtonpost.com/opinions/2021/12 /21/female-crash-test-dummies-nhtsa/.

"Mom Approved: Two Busy Partners Share Why They Think Autonomous Driving Could Help Families and Small Businesses." Waymo, December 2, 2022. https://ltad.com/news/az-mom-squad-approved.html.

Mom, Gijs. *The Electric Vehicle: Technology and Expectations in the Automobile Age*. Baltimore: Johns Hopkins University Press, 2013.

Montgomery, Charlotte. *Handbook for the Woman Driver*. New York: Vanguard Press, 1955.

Moorhouse, H. F. *Driving Ambitions: An Analysis of the American Hot Rod Enthusiasm*. Manchester: Manchester University Press, 1991.

Mumford, Lewis. *The Myth of the Machine: Technics and Human Development*. London: Secker & Warburg, 1967.

Munro, Alice. *Lives of Girls and Women: A Novel*. New York: Knopf Doubleday Publishing Group, 2011.

Nauen, Elinor, ed. *Ladies, Start Your Engines: Women Writers on Cars and the Road*. London: Faber and Faber, 1996.

Neil, Dan. "2023 Cadillac Lyriq: An Electric Leap Forward." *Wall Street Journal*, December 1, 2022. https://www.wsj.com/articles/2023-cadillac -lyriq-an-electric-leap-forward-11669927707?page=1.

"NHTSA's Crash Test Dummies." NHTSA, https://www.nhtsa.gov/ nhtsas-crash-test-dummies.

Nichols, Nancy A. "The Car Office Is the Ultimate Sign of Our Times." *Boston Globe*, February 17, 2021. https://www.bostonglobe.com/2021/02/17 /opinion/car-office-is-ultimate-sign-our-times/.

———. "Cupholders Are Everywhere," *The Atlantic*, April 22, 2018. https://www.theatlantic.com/technology/archive/2018/04/cup holders-are-everywhere/558545/.

———. "Is a College Bumper Sticker Tantamount to Bragging?" *Cognoscenti*, March 26, 2014. https://www.wbur.org/cognoscenti/2014/03/26 /college-bumper-sticker-nancy-a-nichols.

———. "Memoirs of a Used Car Salesman's Daughter." *Creative Nonfiction*, Issue 24.

———ed. *Reach for the Top: Women and the Changing Facts of Work Life.* Cambridge, MA: Harvard Business Review, 1994.

———. "Whatever Happened to Rosie the Riveter?" *Harvard Business Review*, July–August 1993. https://hbr.org/1993/07/whatever-happened-to-rosie -the-riveter.

Nochlin, Linda. *The Body in Pieces: The Fragment as a Metaphor of Modernity.* London: Thames & Hudson, 2001.

Norton, Peter D. *Fighting Traffic: The Dawn of the Motor Age in the American City.* Cambridge, MA: MIT Press, 2011.

———. "Street Rivals: Streetwalking and the Invention of the Motor Age Street." *Technology and Culture* 48, no. 2 (April 2007): 331–59.

———. "Traffic Safety for the Motor Age." In *Fighting Traffic: The Dawn of the Motor Age in the American City*, chap. 8, Cambridge, MA: MIT Press, 2011.

Ogilvy, David. *Ogilvy on Advertising.* New York: Vintage Books, 1985.

Parissien, Steven. *The Life of the Automobile: The Complete History of the Motor Car.* New York: Thomas Dunne Books, 2014.

Parkin, Katherine J. *Women at the Wheel: A Century of Buying, Driving, and Fixing Cars.* Philadelphia: University of Pennsylvania Press, 2017.

Peek, Jeff. "Pony Cars Stamps Will Make It Cool to Write Letters." *Hagerty*, July 18, 2022. https://www.hagerty.com/media/news/pony-cars-stamps -make-it-cool-to-write-letters/.

Perret, Geoffrey. *A Dream of Greatness: The American People, 1945–1963.* New York: Coward, McCann & Geoghegan, 1979.

Pressman, Aaron. "Mass Pike EV Chargers Took the Holiday Weekend Off." *Boston Globe*, May 30, 2023. https://www.bostonglobe.com/2023/05/30 /business/mass-pike-ev-chargers-took-holiday-weekend-off/#:~:text =The%20electric%20vehicle%20chargers%20installed,t%20totally%20 out%20of%20luck.

———. "Massachusetts Needs More EV Chargers—Wicked Badly." *Boston Globe*, August 11, 2023. https://www.bostonglobe.com/2023/08/11 /business/massachusetts-needs-wicked-more-ev-chargers/#:~:text =Massachusetts%20will%20need%20to%20dramatically,according%20 to%20a%20new%20report.

Pynchon, Thomas. *The Crying of Lot 49*. Cutchogue, NY: Buccaneer Books, 1997.

"The Queen's Favourite Cars and How She Earned 'Princess Auto Mechanic' Nickname." *Mirror*, September 10, 2022. https://www.mirror.co.uk/news/royals/queens-favourite-cars-how-earned-27960067.

Rae, John Bell. *The American Automobile: A Brief History*. Chicago: University of Chicago Press, 1965.

Reed, Philip. "Confessions of a Car Salesman." *Edmunds*, 2000.

Rich, Adrienne. *On Lies, Secrets, and Silence: Selected Prose 1966–1978*. New York: W. W. Norton, 1995.

"Ricky Gervais Slams Caitlyn Jenner in Golden Globes Monologue." YouTube, January 10, 2016. https://www.youtube.com/watch?v=86CT53xvOFY.

Rivelli, Elizabeth. "What Is Average Mileage Per Year?" *Car and Driver*, February 24, 2023. https://www.caranddriver.com/auto-loans/a32880477/average-mileage-per-year/#:~:text=The%20data%20also%20shows%20that,for%20females%20is%20only%2010%2C142.

Rothfeder, Jeffrey. *Driving Honda: Inside the World's Most Innovative Car Company*. New York: Penguin Publishing Group, 2015.

"Row over *Vogue Arabia* Cover Featuring Saudi Princess in Driving Seat." *Guardian*, May 31, 2018. https://www.theguardian.com/world/2018/may/31/row-over-vogue-arabia-cover-featuring-saudi-princess-in-driving-seat.

Rutherford, Janice Williams. *Selling Mrs. Consumer: Christine Frederick and the Rise of Household Efficiency*. Athens: University of Georgia Press, 2010.

Sachs, Wolfgang. *For Love of the Automobile: Looking Back into the History of Our Desires*. Berkeley: University of California Press, 1992.

Satariano, Adam. "Meta's Ad Practices Ruled Illegal under E.U. Law." *New York Times*, January 4, 2023. https://www.nytimes.com/2023/01/04/technology/meta-facebook-eu-gdpr.html.

Scharff, Virginia. *Taking the Wheel: Women and the Coming of the Motor Age*. Albuquerque: University of New Mexico Press, 1992.

Schmitt, Angie. *Right of Way: Race, Class, and the Silent Epidemic of Pedestrian Deaths in America*. Washington, DC: Island Press, 2020.

Schwab, Laura. "Life outside the Boys Club: Why I Spoke up about Rivian's Toxic Bro Culture (and Got Fired)." *Medium*, November 4, 2021. https://medium.com/@lauraschwab1/life-outside-the-boys-club-why-i-spoke-up-about-rivians-toxic-bro-culture-and-got-fired-15600b999ae7.

Scott, Victoria. "Your Car Is Tracking You Just as Much as Your Smartphone Is—and Your Data Is at Risk." Drive, June 8, 2022. https://www.thedrive.com/news/think-your-smartphone-is-tracking-you-your-cars-doing-it-too.

Sehl, Katie. "The Bizarre History of Trying to Design Cars for Women."
    VICE, April 23, 2019. https://www.vice.com/en/article/43jvvb/cars
    -for-women-design-history-volvo.

Sen, Gautam. *The Automobile: An Indian Love Affair*. Delhi: Penguin Random
    House India, 2021.

Serna, Joseph, and Veronica Rocha. "Bruce Jenner Fights Wrongful-Death
    Lawsuit in Fatal Malibu Crash." *Los Angeles Times*, May 28, 2015. https
    ://www.latimes.com/local/lanow/la-me-ln-bruce-jenner-lawsuit-in-fatal
    -malibu-crash-20150528-story.html.

Seymour, Miranda. *The Bugatti Queen*. New York: Simon & Schuster, 2005.

Shankman, Sabrina, and Aaron Pressman. "We Drove around New England
    Looking for EV Chargers: It Was Better than Expected." *Boston Globe*,
    January 10, 2023. https://www.bostonglobe.com/2023/01/10/science
    /we-drove-around-new-england-looking-ev-chargers-best-doughnuts-it
    -was-easier-than-we-expected/.

Shill, Gregory H. "Car Crashes Aren't Always Unavoidable." *Atlantic*,
    July 9, 2019. https://www.theatlantic.com/ideas/archive/2019/07/car
    -crashes-arent-always-unavoidable/592447/.

———. "How Vehicular Intimidation Became the Norm." *Atlantic*,
    November 3, 2020. https://www.theatlantic.com/ideas/archive/2020/11
    /how-trump-train-trucks-became-a-political-weapon/616979/.

———. "Should Law Subsidize Driving?" *New York University Law Review* 95,
    no. 2 (May 2020): 500–579.

Shoup, Donald. *Parking and the City*. London: Taylor & Francis, 2018.

Shrivastava, Jayawati, ed. *Lady Driver: Stories of Women behind the Wheel*.
    New Delhi: Zubaan, 2017.

Sloan, Alfred Pritchard. *My Years with General Motors*. New York: Doubleday, 1964.

Smiley, Lauren. "'I'm the Operator': The Aftermath of a Self-Driving Tragedy."
    Wired, March 8, 2022. https://www.wired.com/story/uber-self-driving
    -car-fatal-crash/.

Smith, Constance A. *Damsels in Design: Women Pioneers in the Automotive
    Industry, 1939—1959*. Atglen, PA: Schiffer Publishing, 2018.

Smith, Woodruff D. *Consumption and the Making of Respectability, 1600–1800*.
    London: Routledge, 2002.

Smith, Zadie. *Changing My Mind: Occasional Essays*. New York: Penguin
    Books, 2009.

Sorin, Gretchen Sullivan. *Driving While Black: African American Travel and the
    Road to Civil Rights*. New York: W. W. Norton, 2020.

Sottile, Zoe. "You Will Soon Be Able to Jaywalk Ticket-Free in California." CNN, October 2, 2022. https://www.cnn.com/2022/10/02/us/california -jaywalking-law-trnd.

Spellman, Susan V. "All the Comforts of Home: The Domestication of the Service Station History, 1920–1940." *Journal of Popular Culture* 37, no. 3 (2004): 463–77. https://onlinelibrary.wiley.com/doi/10.1111/j.0022 -3840.2004.00080.x.

Standage, Tom. *A Brief History of Motion: From the Wheel, to the Car, to What Comes Next.* London: Bloomsbury Publishing, 2021.

Studebaker Corporation. *Recruiting, Selecting and Training Women for Automotive Maintenance Service: A Manual for Dealers and Service Men in the Automotive Maintenance Industry Who Are Supporting the War Effort by "Keeping 'em Rolling."* Sound Bend, IN: Studebaker Corporation, 1943.

Tarkington, Booth. *The Magnificent Ambersons.* New York: Modern Library, 1998.

Taylor, Alex. "IACOCCA'S MINIVAN How Chrysler Succeeded in Creating One of the Most Profitable Products of the Decade." *Fortune*, May 30, 1994. https://money.cnn.com/magazines/fortune/fortune_archive/1994 /05/30/79354/.

Tengler, Steve. "New 'Pink Tax' Study Shows Women Pay Upwards of $7,800 More for Car Ownership." *Forbes*, October 27, 2021. https://www.forbes .com/sites/stevetengler/2021/10/27/new-pink-tax-study-shows-women -pay-upwards-of-7800-more-for-car-ownership/?sh=52b6272463d.

Tenis, Fatma. "Saudi Activist Who Led Campaign to Legalize Driving for Women Is Released from Jail." NPR, February 10, 2021. https://www.npr .org/2021/02/10/966258281/loujain-al-hathloul-saudi-activist-jailed -for-driving-has-been-released.

Terkel, Studs. *Working: People Talk about What They Do All Day and How They Feel about What They Do.* New York: New Press, 2011.

Tichi, Cecelia. *Shifting Gears: Technology, Literature, Culture in Modernist America.* Chapel Hill: University of North Carolina Press, 1987.

Tigue, Kristoffer. "Heavy Electric Vehicles Jeopardize Climate Action and Public Safety, Experts Warn." Inside Climate News, February 7, 2023. https://insideclimatenews.org/todaysclimate/heavy-electric-vehicles -jeopardize-climate-action-and-public-safety-experts-warn/.

"Times Have Changed and So Have Clothes and Cars: Convertible Coupé and Suits." *Good Housekeeping* 11 (1940): 48–59.

Torchinsky, Jason. *Robot, Take the Wheel: The Road to Autonomous Cars and the Lost Art of Driving.* New York: Apollo Publishers, 2019.

Townsend, Anthony M. *Ghost Road: Beyond the Driverless Car*. New York: W. W. Norton, 2020.

"'Trump Train' Ambushes Biden-Harris Campaign Bus in Texas." YouTube, October 31, 2020. https://youtu.be/i0jglSIgRbM.

Twain, Mark, Warner, Charles Dudley. *The Gilded Age: A Tale of Today*. New York: Harper & Brothers Publishers, 1904.

Valdes-Dapena, Peter. "Sony and Honda Reveal Their New Car Brand." CNN, January 5, 2023. https://www.cnn.com/2023/01/04/tech/sony -honda-afeela/index.html.

Van der Kolk, Bessel A. *The Body Keeps the Score: Brain, Mind, and Body in the Healing of Trauma*. London: Penguin Publishing Group, 2014.

Vanderbilt, Tom. *Traffic: Why We Drive the Way We Do (and What It Says about Us)*. New York: Vintage Books, 2009.

Veblen, Thorstein. *The Theory of the Leisure Class*. Oxford: Oxford University Press, 2009.

"Vehicle-to-Everything." Wikipedia. https://en.wikipedia.org/wiki/Vehicle -to-everything.

Walsh, Margaret. "Gender and Automobility: The Pioneering and Early Years." *Automobile in American Life and Society*. http://www.autolife.umd.umich .edu/Gender/Walsh/G_Overview2.htm.

———. "Gender and the Automobile in the United States." *Automobile in American Life and Society*. http://www.autolife.umd.umich.edu/Gender /Walsh/G_Overview.htm.

———. "Gendering Mobility: Women, Work and Automobility in the United States." *History* 93, no. 3 (July 2008): 376–95. www.jstor.org /stable/24428395.

Ward, Hilda. *The Girl and the Motor*. Cincinnati: Gas Engine Publishing Co., 1908.

"'We Are No Longer the Customers': Inside the 'Age of Surveillance Capitalism.'" WBUR, January 15, 2019. https://www.wbur.org/onpoint/2019/01/15 /surveillance-capitalism-online-data-privacy-facebook-google-amazon.

Weart, Kimi. *Cars! Cars! Cars! Featuring Cars from the Collection of the Museum of Modern Art*. New York: Museum of Modern Art, 2021.

Welles, Orson, Peter Bogdanovich, and Jonathan Rosenbaum. *This Is Orson Welles*. New York: Hachette Books, 1998.

Wells, Christopher W. *Car Country: An Environmental History*. Seattle: University of Washington Press, 2014.

Wharton, Edith. *The Custom of the Country*. New York: Charles Scribner's Sons, 1913.

Wharton, Edith. *A Motor-flight through France*. New York: C. Scribner's Sons, 1908.

"When Queen Elizabeth Took King Abdullah for a Drive in Her Land Rover." *PBS NewsHour*, January 23, 2015. https://www.pbs.org/newshour/world/driving-king-anecdote-goes-viral.

"Why the Man Always Drives." *Week*, January 11, 2015. https://theweek.com/articles/496642/why-man-always-drives.

"Why Women Need the Office." *Economist*, August 28, 2021. https://www.economist.com/business/2021/08/28/why-women-need-the-office.

Whyte, Kenneth. *The Sack of Detroit: General Motors, Its Enemies, and the End of American Enterprise*. New York: Alfred A. Knopf, 2021.

Wilson, Emily R., trans. *The Odyssey*. New York: W. W. Norton, 2018.

Winchester, Simon. *The Perfectionists: How Precision Engineers Created the Modern World*. New York: HarperCollins, 2018.

Wolfe, Tom. *The Kandy-Kolored Tangerine-Flake Streamline Baby*. New York: Picador, 2009.

Wollen, Peter, and Joe Kerr, eds. *Autopia: Cars and Culture*. London: Reaktion Books, 2002.

Womack, James P., Daniel T. Jones, and Daniel Roos. *The Machine That Changed the World: Based on the Massachusetts Institute of Technology 5-Million Dollar 5-Year Study on the Future of the Automobile*. New York: Rawson Associates, 1990.

Wright, Priscilla Hovey. *The Car Belongs to Mother*. New York: Riverside Press, 1939.

Yeager, Robert C. "More Women Muscle in on the World of Vintage Cars." *New York Times*, August 11, 2021. https://www.nytimes.com/2021/08/11/business/women-car-collecting.html.

Zdanowicz, Christina, Claudia Dominguez, Lauren M. Johnson, and Sara Smart. "The Waukesha Victims Included an 8-Year-Old Boy, A Loving Grandmother and a Woman Excited to Make Her Debut in the Dancing Grannies." CNN, October 26, 2022. https://www.cnn.com/2021/11/22/us/waukesha-wisconsin-parade-victims/index.html.

Zhu, Motato, Songzhu Zhao, Jeffrey H. Coben, and Gordon S. Smith. "Why More Male Pedestrians Die in Vehicle-Pedestrian Collisions than Female Pedestrians: A Decompositional Analysis." *Injury Prevention* 19 (2013): 227–31. doi: 10.1136/injuryprev-2012-040594.

Zoellner, Tom. *The National Road: Dispatches from a Changing America*. Berkeley, CA: Counterpoint Press, 2020.

Zuboff, Shoshana. *The Age of Surveillance Capitalism: The Fight for a Human Future at the New Frontier of Power*. New York: PublicAffairs, 2019.

Notes

Chapter 1

1   Mr. Jean François Bouzanquet, quoted in Robert C. Yeager, "More Women Muscle in on the World of Vintage Cars," *New York Times*, August 11, 2021, https://www.nytimes.com/2021/08/11/business /women-car-collecting.html; Jean-François Bouzanquet, *Fast Ladies: Female Racing Drivers 1888–1970* (Poundbury, UK: Veloce Publishing, 2009).

2   Virginia Scharff, *Taking the Wheel: Women and the Coming of the Motor Age* (Albuquerque: University of New Mexico Press, 1992), 10.

3   Sinclair Lewis, *Babbitt* (New York: Bantam Dell, 1998), 2.

4   Walter A. Friedman, *Birth of a Salesman: The Transformation of Selling in America* (Cambridge, MA: Harvard University Press, 2004), 269.

5   Friedman, *Birth of a Salesman*, 10.

6   Bruce Barton, *The Man Nobody Knows: A Discovery of Jesus* (Indianapolis: Bobbs-Merrill, 1925), 104.

7   "Honesty/Ethics in Professions," Gallup, https://news.gallup.com/poll /1654/honesty-ethics-professions.aspx.

8   James J. Flink, *The Car Culture* (Cambridge, MA: MIT Press, 1976), 192.

9   Chris Lezotte, "What Would Miss Daisy Drive? The Road Trip Film, the Automobile, and the Woman behind the Wheel," *Journal of Popular Culture* 54, no. 5 (October 2021): 987–1011, https://doi.org /10.1111/jpcu.13061.

10   Friedman, *Birth of a Salesman*, 214

11   Cynthia Wright, "'Feminine Trifles of Vast Importance': Writing
        Gender into the History of Consumption." In *Gender Conflicts: New
        Essays in Women's History*, ed. Franca Iacovetta and Mariana Valverde
        (Toronto: University of Toronto Press, 1992), 229–260. http://www
        .jstor.org/stable/10.3138/j.ctt2ttstj.11.

12   Franca Iacovetta and Mariana Valverde, eds., *Gender Conflicts: New
        Essays in Women's History* (Toronto: University of Toronto Press, 1992).

13   Friedman, *Birth of a Salesman*, 222

14   Robert Genat, *The American Car Dealership* (St. Paul, MN: MBI
        Publishing, 1999), 103.

15   Katherine J. Parkin, *Women at the Wheel: A Century of Buying, Driving,
        and Fixing Cars* (Philadelphia: University of Pennsylvania Press,
        2017), 35.

16   Genat, *American Car Dealership*, 93.

Chapter 2

1   Katherine J. Parkin, *Women at the Wheel: A Century of Buying, Driving,
        and Fixing Cars* (Philadelphia: University of Pennsylvania Press,
        2017), 15.

2   David L. Lewis and Laurence Goldstein, eds., *The Automobile and
        American Culture* (Ann Arbor: University of Michigan Press, 1983), 127.

3   F. Scott Fitzgerald, *The Great Gatsby* (Wisehouse Classics Edition,
        1925), 64.

4   Virginia Scharff, *Taking the Wheel: Women and the Coming of the Motor
        Age*. (Albuquerque: University of New Mexico Press, 1992), 21.

5   Edith Wharton, *The Custom of the Country*. (New York: Charles
        Scribner's Sons, 1913), 116.

6   Beth Bailey, *From Front Porch to Back Seat: Courtship in Twentieth-
        Century America*. (Baltimore: Johns Hopkins University Press,
        1989), 86.

7   Bailey, *From Front Porch to Back Seat: Courtship in Twentieth-Century
        America*, 86.

8   Bailey, *From Front Porch to Back Seat: Courtship in Twentieth-Century
        America*, 3.

9   David L. Lewis, "From Rumble Seats to Rockin' Van," in *The
        Automobile and American Culture*, ed. David L. Lewis and Laurence
        Goldstein (Ann Arbor: University of Michigan Press, 1983), 123–33.

10   Bailey, *From Front Porch to Back Seat: Courtship in Twentieth-Century America*, 23.

11   Virginia Scharff, *Taking the Wheel: Women and the Coming of the Motor Age*, 140.

12   *Waukegan News Sun*, June 5, 1992.

13   Lewis, "From Rumble Seats to Rockin' Van."

Chapter 3

1   Thirty-three men from Waukegan, Illinois, are listed on the Vietnam Veterans Memorial in Washington, DC. That number represents just over 10 percent of the total number of men from Illinois who died in the war.

2   Margaret Walsh, "Gender and Automobility: The Pioneering and Early Years," *Automobile in American Life and Society*, http://www .autolife.umd.umich.edu/Gender/Walsh/G_Overview2.htm.

3   Tussy Advertisement, *Seventeen* magazine, September 1966.

4   *Automobile Garments and Requisites: Imported and Domestic Models for Men and Women* (New York: Saks and Company, 1904), 206. In the collection of the New York Public Library.

5   Mary Walker Harper, "The Woman Who Drives a Car Should Know How to Buy and Run One," *Ladies' Home Journal*, September 1915.

6   Harper, "Woman Who Drives a Car Should Know How to Buy and Run One."

7   Dorothy Levitt, *The Woman and the Car: A Chatty Little Handbook for All Women Who Motor or Who Want to Motor* (London: John Lane, 1909).

8   Edith Wharton. *A Motor-flight Through France*. (New York: C. Scribner's Sons, 1908), 1.

9   Virginia Scharff, *Taking the Wheel: Women and the Coming of the Motor Age* (Albuquerque: University of New Mexico Press, 1992), 16.

10   Michael Berger, "Women Drivers!: The Emergence of Folklore and Stereotypic Opinions Concerning Feminine Automotive Behavior," *Women's Studies International Forum* 9, no. 3 (1986): 257–63.

11   Berger, "Women Drivers!"

12   F. Scott Fitzgerald, *The Great Gatsby* (Wisehouse Classics Edition, 1925).

13   Deborah Clarke, *Driving Women: Fiction and Automobile Culture in Twentieth-Century America* (Baltimore: Johns Hopkins University Press, 2007), 19.

14   Vince Guerrieri, "Jordan Motor Car Co.'s Famous Ad," *Ohio Magazine*, February 2018, https://www.ohiomagazine.com/ohio-life /article/driving-inspiration.

15   James J. Flink, "The Three Stages of American Automobile Consciousness," *American Quarterly* 24, no. 4 (October 1972): 451–73.

16   Roland Marchand, *Advertising the American Dream: Making Way for Modernity, 1920–1940* (Berkeley: University of California Press, 1985), 136.

17   Marchand, *Advertising the American Dream*, 136.

18   Marchand, *Advertising the American Dream*, 139.

19   John Keats, *The Insolent Chariots* (Philadelphia: Lippincott, 1958), 36. Quote attributed to Mr. George William Walker, vice president in charge of styling at Ford.

20   Copy from an image from the Schlesinger Library at Harvard, JPEG 1315353, Harold Constain, photographer.

21   "Times Have Changed and So Have Clothes and Cars: Convertible Coupé and Suits," *Good Housekeeping* 11 (1940): 48–59.

22   *Vagina dentata* is an old myth in which a woman's vagina was said to contain teeth to emasculate a man. Wikipedia, https://en.wikipedia .org/wiki/Vagina_dentata.

23   Keats, *Insolent Chariots*.

24   Laura L. Behling, "'The Woman at the Wheel': Marketing Ideal Womanhood, 1915–1934," *Journal of American Culture* 20, no. 3 (Fall 1997): 13–30.

25   Behling, "'The Woman at the Wheel': Marketing Ideal Womanhood, 1915–1934," 17.

26   Beth Bailey, *From Front Porch to Back Seat: Courtship in Twentieth Century America.* (Baltimore: Johns Hopkins University Press, 1988), 70.

27   Ed Heys, "Cadillacs Crown Jewels," *Hemmings*, September 23, 2018, https://www.hemmings.com/stories/article/cadillacs-crown-jewels.

28   Robert A. Fria, *Mustang Genesis: The Creation of the Pony Car* (Jefferson, NC: McFarland & Company, 2010), 66.

29   Fria, *Mustang Genesis*, 83.

30   Fria, *Mustang Genesis*, 171.

31   Clarke, *Driving Women: Fiction and Automobile Culture in Twentieth-Century America*, 18.

32   See https://www.youtube.com/watch?v=WCg7uesivkY.

33   Katherine J. Parkin, *Women at the Wheel: A Century of Buying, Driving,
        and Fixing Cars* (Philadelphia: University of Pennsylvania Press,
        2017), 57.

34   Lauren Michele Jackson, "The Zora Neal Hurston We Don't Talk
        About," *New Yorker*, February 14, 2002, https://www.newyorker.com
        /books/page-turner/the-zora-neale-hurston-we-dont-talk-about.

35   Jeffrey Rothfeder. *Driving Honda: Inside the World's Most Innovative
        Car Company.* (New York: Penguin Publishing Group, 2014), 14.

36   Rothfeder. *Driving Honda, Inside the World's Most Innovative
        Company*, 12.

Chapter 4

1   Daniel Guillory, "Bel Air: The Automobile as Art Object," *Michigan
        Quarterly Review* 19, no. 4 (Fall 1980): 663. https://quod.lib.umich
        .edu/m/mqrarchive/act2080.0019.004?xc=1&g=mqrg&node=act2080
        .0019.004:31&view=image&seq=239&size=100.

2   "Spuot Etymology," Etymology, https://etymologeek.com/gmh/spuot.

3   Lauren Elkin, *Flâneuse: Women Walk the City in Paris, New York, Tokyo,
        Venice, and London* (New York: Farrar, Straus and Giroux, 2017).

4   Virginia Scharff, *Taking the Wheel: Women and The Coming of the Motor
        Age*, 13.

5   Michael Berger, "Women Drivers!: The Emergence of Folklore and
        Stereotypic Opinions Concerning Feminine Automotive Behavior,"
        Women's Studies International Forum 9, no. 3 (1986): 257–63.

6   Georgine Clarsen, *Eat My Dust: Early Women Motorists* (Baltimore:
        Johns Hopkins University Press, 2008), 101.

7   *Cadillac Craftsman*, July 3, 1929.

8   Georgine Clarsen, "The 'Dainty Female Toe' and the 'Brawny Male
        Arm': Conceptions of Bodies and Power in Automobile Technology,"
        Australian Feminist Studies 15, no. 32 (2000): 153–63, https://doi.org
        /10.1080/08164640050138671.

9   Tom McCarthy, *Auto Mania: Cars, Consumers, and the Environment*
        (New Haven, CT: Yale University Press, 2007), 7.

10   Deborah Clarke, *Driving Women: Fiction and Automobile Culture in
        Twentieth-Century America* (Baltimore: Johns Hopkins University
        Press), 119–20.

11   Clarke, *Driving Women*, 45.

12   McCarthy, *Auto Mania*, 6–7.

13   Gijs Mom, *The Electric Vehicle: Technology and Expectations in the Automobile Age* (Baltimore: Johns Hopkins University Press, 2013), 59.

14   Margaret Walsh, "Gender and the Automobile in the United States," *Automobile in American Life and Society*, http://www.autolife.umd. umich.edu/Gender/Walsh/G_Overview.htm.

15   "Columbia Electric Key Facts," National Motor Museum, https ://nationalmotormuseum.org.uk/vehicle-collection/columbia-electric/.

16   Mom, *Electric Vehicle*, 58.

17   Miranda Seymour, *The Bugatti Queen* (New York: Simon & Schuster, 2005), 81.

18   Seymour, *Bugatti Queen*, 147.

Chapter 5

1   Adrienne Rich, *On Lies, Secrets, and Silence: Selected Prose 1966–1978* (New York: W. W. Norton, 1995).

2   Bessel A. Van der Kolk, *The Body Keeps the Score: Brain, Mind, and Body in the Healing of Trauma* (London: Penguin Publishing Group, 2015), 221.

3   Georgine Clarsen, "The 'Dainty Female Toe' and the 'Brawny Male Arm': Conceptions of Bodies and Power in Automobile Technology," *Australian Feminist Studies* 15, no. 32 (2000): 153–63, https://doi.org /10.1080/08164640050138671.

4   Reyner Banham, *Los Angeles: The Architecture of Four Ecologies*, University of California Press, 2009, 196.

5   Joan Didion, *Play It as It Lays* (New York: Farrar, Strauss and Giroux, 2005), 17.

6   Laura L. Behling, "'The Woman at the Wheel': Marketing Ideal Woman-hood, 1915–1934," *Journal of American Culture* 20, no. 3 (Fall 1997): 28.

7   Katherine J. Parkin, *Women at the Wheel: A Century of Buying, Driving, and Fixing Cars* (Philadelphia: University of Pennsylvania Press, 2017), 21.

8   Parkin, *Women at the Wheel*, 31.

Chapter 6

1   Deborah Clarke, *Driving Women: Fiction and Automobile Culture in Twentieth-Century America* (Baltimore: Johns Hopkins University Press, 2007), 4.

2   Clarke, *Driving Women*, 108.

3   "Why the Man Always Drives," *Week*, January 11, 2015, https://theweek .com/articles/496642/why-man-always-drives; see also Stephen J. Dubner,

"Are Women Being Taken for a Ride?" Freakanomics, January 11, 2010, https://freakonomics.com/2010/01/are-women-being-taken-for-a-ride/.

4   Matthew B. Crawford, *Why We Drive: Toward a Philosophy of the Open Road* (Boston: Mariner Books, 2020), 140.

5   Crawford, *Why We Drive*, 140.

6   Andrea Hiott, *Thinking Small: The Long, Strange Trip of the Volkswagen Beetle* (New York: Ballantine Books, 2012), 53.

7   Hiott, *Thinking Small*, 362.

8   Hiott, *Thinking Small*, 351.

9   "History of the Beetle's Bud Vase," courtesy of the VW newsroom, https://www.timmonsvw.com/the-history-of-the-beetles-bud-vase/.

10   Parkin, interviewed by Thomas Carroll, "Women, Cars and Liberation," *Monmouth*, Fall 2017, https://www.monmouth.edu/magazine/women-cars-liberation/.

11   David L. Lewis and Laurence Goldstein, eds., *The Automobile and American Culture* (Ann Arbor: University of Michigan Press, 1983), 142.

12   Leo Marx, *The Machine in the Garden: Technology and the Pastoral Ideal in America* (Kiribati: Oxford University Press, 2000), 343.

13   Beth L. Bailey, *From Front Porch to Back Seat: Courtship in Twentieth-Century America* (Baltimore: Johns Hopkins University Press, 1989), 98.

Chapter 7

1   Tom Vanderbilt, *Traffic: Why We Drive the Way We Do (and What It Says about Us)* (New York: Vintage Books, 2009), 218.

2   Adam Gopnik, "Jane Jacob's Street Smarts," *New Yorker*, September 26, 2016, https://www.newyorker.com/magazine/2016/09/26/jane-jacobs-street-smarts.

3   Peter D. Norton, *Fighting Traffic: The Dawn of the Motor Age in the American City* (Ukraine: MIT Press, 2011), 66.

4   Peter D. Norton, "Street Rivals: Jaywalking and the Invention of the Motor Age Street," *Technology and Culture* 48, no. 2 (April 2007): 331–59.

5   Norton, *Fighting Traffic: The Dawn of the Motor Age in the American City*, 333.

6   Gregory H. Shill, "Should Law Subsidize Driving?" *New York University Law Review* 95, no. 2 (May 2020): 527.

7   Orson Welles, Peter Bogdanovich, and Jonathan Rosenbaum, *This Is Orson Welles* (New York: Hachette Books, 1998).

8   Charles Dickens, *A Tale of Two Cities* (New York: Penguin Books, 2007),
         116, https://www.google.com/books/edition/A_Tale_of_Two_Cities
         /j_5UbgB_9isC?hl=en&gbpv=1&bsq=ever%20in%20the%20way.

9   Norton, *Fighting Traffic: The Dawn of the Motor Age in the American
         City.*

10   Norton, *Fighting Traffic,* 72.

11   Norton, *Fighting Traffic,* 75.

12   Peter D. Norton, "Traffic Safety for the Motor Age," in *Fighting
         Traffic: The Dawn of the Motor Age in the American City* (Cambridge,
         MA: MIT Press, 2011), 221.

13   Norton, "Traffic Safety for the Motor Age," 221.

14   Motato Zhu et al. "Why More Male Pedestrians Die in Vehicle-
         Pedestrian Collisions than Female Pedestrians: A Decompositional
         Analysis," *Injury Prevention* 19 (2013): 227–31, doi: 10.1136/injuryprev
         -2012-040594.

15   Gregory H. Shill, "Car Crashes Aren't Always Unavoidable," *Atlantic,*
         July 9, 2019, https://www.theatlantic.com/ideas/archive/2019/07/car
         -crashes-arent-always-unavoidable/592447/.

16   Ravi Mangla, "The Secret History of Jaywalking: The Disturbing
         Reason It Was Outlawed—and Why We Should Lift the Ban," *Salon,*
         August 20, 2015, https://www.salon.com/2015/08/20/the_secret
         _history_of_jaywalking_the_disturbing_reason_it_was_outlawed
         _and_why_we_should_lift_the_ban/.

17   Shill, "Should Law Subsidize Driving?," 530.

18   Leo Marx, *The Machine in the Garden: Technology and the Pastoral Ideal
         in America.* (Kiribati: Oxford University Press, 2000), 365.

19   Zoe Sottile, "You Will Soon Be Able to Jaywalk Ticket-Free in
         California," CNN, October 2, 2022, https://www.cnn.com/2022
         /10/02/us/california-jaywalking-law-trnd.

20   Czeslaw Milosz, "A Magic Mountain," *The Collected Poems 1931–1987*
         (The Ecco Press, 1988), 317.

21   Donald Shoup, *The High Cost of Free Parking* (New York: Routledge,
         2017), 591.

22   Shoup, *The High Cost of Free Parking,* xx.

23   Katherine Parkin, *Women at the Wheel: A Century of Buying, Driving, and
         Fixing Cars* (Philadelphia: University of Pennsylvania Press, 2017), 74.

24   Shoup, *The High Cost of Free Parking,* xx.

25   *Highland Park Press,* October 15, 1932.

Chapter 8

1   Dan Neil, "Honda Odyssey: The Gas-Powered Marital Aid," *Wall Street Journal*, August 10, 2017.

2   Michael Laris, "Minivans Are the Future of Transportation—Just Don't Call Them Minivans," *Washington Post*, September 28, 2018.

3   Chris Lezotte, "The Evolution of the 'Chick Car' Or: What Came First, the Chick or the Car?" *The Journal of Popular Culture* 45, no. 3 (2012): 516.

4   Lezotte, 516.

5   Nancy A. Nichols, "Whatever Happened to Rosie the Riveter?" *Harvard Business Review*, July–August 1993, https://hbr.org/1993/07/whatever-happened-to-rosie-the-riveter.

6   Ruth Schwartz Cowan. *More Work for Mother: The Ironies of Household Technology from the Open Hearth to the Microwave* (New York: Basic Books, 1983), 85.

7   Tom Vanderbilt, *Traffic: Why We Drive the Way We Do (and What It Says about Us)* (New York: Vintage Books, 2009), 135.

8   Martha N. Beck, *Breaking Point: Why Women Fall Apart and how They Can Re-create Their Lives.* (New York: Times Books, 1997), 15.

9   Ann Crittenden, *The Price of Motherhood: Why the Most Important Job in the World is Still the Least Valued.* (New York: Henry Holt and Company, 2002), 22.

10   Frank Ahrens, "A Tankful of Suburban Angst," *Washington Post*, March 26, 1998, https://www.washingtonpost.com/archive/lifestyle/1998/03/26/a-tankful-of-suburban-angst/64a4299b-da73-4da1-8f41-3b6d5e877a01/.

11   Neil, "Honda Odyssey: The Gas-Powered Marital Aid," *Wall Street Journal*.

12   Glennon Doyle, *Love Warrior: A Memoir* (New York: Flatiron Books, 2017), 130.

13   Nancy A. Nichols, "Is a College Bumper Sticker Tantamount to Bragging?" *Cognoscenti*, March 26, 2014, https://www.wbur.org/cognoscenti/2014/03/26/college-bumper-sticker-nancy-a-nichols.

14   E. M. Forster, "The Machine Stops," *Oxford and Cambridge Review*, November 1909, 15.

Chapter 9

1   James J. Flink, *The Car Culture* (Cambridge, MA: MIT Press, 1987), 193.

2   Flink, *Car Culture*, 193.

3   Tom McCarthy, "The Black Box in the Garden: Consumers and the
        Environment," in *A Companion to American Environmental History*,
        ed. Douglas Cazaux Sackman (Hoboken, NJ: Wiley, 2010), 304–24,
        https://doi.org/10.1002/9781444323610.ch16.

4   Flink, *Car Culture*, 197.

5   Flink, *Car Culture*, 197.

6   Tom McCarthy, "The Black Box in the Garden: Consumers and the
        Environment," in *A Companion to American Environmental History*,
        ed. Douglas Cazaux Sackman (Hoboken, NJ: Wiley, 2010), 304–24,
        https://doi.org/10.1002/9781444323610.ch16.

7   Lendol Calder, *Financing the American Dream: A Cultural History of
        Consumer Credit* (Princeton, NJ: Princeton University Press, 1999), 243.

8   Wolfgang Sachs, *For the Love of the Automobile: Looking Back into the
        History of Our Desires* (Berkeley: University of California Press, 1992), 65.

9   Sachs, *For the Love of the Automobile*, 38.

10   Is this Girl Smarter than a Million Men? Ad for the 1924
        Chevrolet Superior Series F. Housed in the collection of the GM
        Heritage Center, Sterling Heights, Michigan. Image number
        DN706-Chev-24-139.

11   "Buy Now, Pay Later: A History of Personal Credit," Harvard
        Business School Historical Collections, https://www.library.hbs.edu
        /hc/credit/credit4d.html.

12   "Ford Motor Credit Marks 50th Anniversary," *F & I and Showroom*,
        August 20, 2009, https://www.fi-magazine.com/317887/ford-motor
        -credit-marks-50th-anniversary.

13   Calder, *Financing the American Dream*, 199.

14   "Americans are Borrowing at Record Levels to Pay for Their Expensive
        Cars," CNN Business, August 25, 2022, https://www.cnn.com/2022
        /08/25/cars/car-price-borrowing/index.html.

15   Mark Twain and Charles Warner. *The Gilded Age: A Tale of Today* (New
        York: Harper & Brothers Publishers, 1904), 292.

16   Sinclair Lewis, *Babbitt* (Czechia: New American Library, 1961), 74.

17   John Bell Rae, *The American Automobile: A Brief History* (Chicago:
        University of Chicago Press, 1965), 215.

18   Angie Schmitt, *Right of Way: Race, Class, and the Silent Epidemic of
        Pedestrian Deaths in America* (Washington, DC: Island Press, 2020), 96.

19   Elizabeth Rivelli, "What Is Average Mileage Per Year?" *Car and
        Driver*, February 24, 2023, https://www.caranddriver.com/auto-loans

/a32880477/average-mileage-per-year/#:~:text=The%20data%20
also%20shows%20that,for%20females%20is%20only%2010%2C142.

20   Keith Bradsher, *High and Mighty: The Dangerous Rise of the SUV* (New
        York: PublicAffairs, 2003).

21   Roland Barthes, *Mythologies* (New York: Hill & Wang, 1957).

22   Roland Barthes, "A Cruel Country," *New Yorker*, September 6, 2010,
        https://www.newyorker.com/magazine/2010/09/13/a-cruel-country.

23   Aurimas Grinys, "The Citroën DS: A Goddess Ahead of Its Time,"
        Dyler, March 21, 2022, https://dyler.com/posts/105/the-citroen-ds-a
        -goddess-ahead-of-its-time.

24   "The DS by Roland Barthes," Citroënët, http://www.citroenet.org.uk
        /passenger-cars/michelin/ds/32.html.

25   Sachs, *For the Love of the Automobile*, 91.

26   Alice Munro. *Lives of Girls and Women: A Novel* (New York: Knopf
        Doubleday Publishing Group, 2011).

27   For a fascinating history of lead in gasoline, see Jamie Kitman, "A
        Brief History of Gasoline: Searching for the Magic Bullet," *Jalopnik*,
        January 28, 2022, https://jalopnik.com/a-brief-history-of-gasoline
        -searching-for-the-magic-bu-1848438134.

28   Nick Kurczewski, "The Science of the New-Car Smell," *Car and
        Driver*, July 10, 2021, https://www.caranddriver.com/features/a369
        70626/science-new-car-smell/.

Chapter 10

1   Charles L. Sanford, "Woman's Place in American Car Culture," from *The
        Automobile in American Culture*, ed. David L. Lewis and Laurence Goldstein,
        University of Michigan Press (1983), Ann Arbor, Michigan. 138.

2   This is from promotional copy provided by the manufacturer. See "A
        Heritage of Heroes, 1940–1949," Jeep, n.d., https://www.jeep.com
        /history/1940s.html.

3   Gary Gastelu, "Jeep Named Most Patriotic Automaker and Brand,"
        Fox News, July 4, 2022, https://www.foxnews.com/auto/jeep-most
        -patriotic-automaker-brand.

4   Joel W. Eastman, *Styling vs. Safety: The American Automobile Industry
        and the Development of Automotive Safety, 1900–1966* (Lanham, MD:
        University Press of America, 1984).

5   For more, see *"Baboon and Young,"* Wikipedia, https://en.wikipedia.
        org/wiki/Baboon_and_Young.

6   "This exhibition addresses the conflicted feelings—compulsion,
        fixation, desire, and rage—that developed in response to cars and car
        culture in the 20th century." See *Automania*, MoMA, https://www
        .moma.org/calendar/exhibitions/5210.

7   Ernest Dichter, *The Strategy of Desire* (London: Taylor & Francis, 2017).

8   Katherine Parkin, *Women at the Wheel: A Century of Buying, Driving, and
        Fixing Cars* (Philadelphia: University of Pennsylvania Press, 2017), 41.

9   *A Coach for Cinderella*, 1932, Fisher Auto Body Ad, Collection of the
        GM Heritage Center, Sterling Heights, Michigan. Image 137294.

10   William Knoedelseder, *Fins: Harley Earl, the Rise of General Motors,
        and the Glory Days of Detroit* (New York: Harper Business, 2019), 242.

11   Knoedelseder, *Fins*, 243.

12   Constance Smith, *Damsels in Design: Women Pioneers in the Automotive
        Industry, 1939–1959* (Atglen, PA: Schiffer Publishing, 2018), 115.

13   Knoedelseder, *Fins*, 244.

14   Parkin, *Women at the Wheel*, 40.

15   Chester Dawson, "Is It Possible to Have Too Many Cup Holders?
        Auto Makers Are Trying to Find Out," *Wall Street Journal*, February 15,
        2018, https://www.wsj.com/articles/need-a-place-to-park-your
        -drink-the-worlds-auto-makers-are-on-it-1518710790.

16   Malcolm Gladwell, "Big and Bad: How the S.U.V. Ran Over
        Automotive Safety," *New Yorker*, January 12, 2004, 28.

17   Sam Dean, "The History of the Car Cup Holder," *Bon Appétit*,
        February 18, 2013, https://www.bonappetit.com/trends/article/the
        -history-of-the-car-cup-holder.

18   Dean, "The History of the Car Cup Holder."

19   "Liebeck v. McDonald's Restaurants," Wikipedia, https://en.wiki
        pedia.org/wiki/Liebeck_v._McDonald%27s_Restaurants.

20   Dan Albert, *Are We There Yet? The American Automobile Past, Present,
        and Driverless* (New York: W. W. Norton, 2019).

Chapter 11

1   "The Queen's Favourite Cars and How She Earned 'Princess Auto
        Mechanic' Nickname," *Mirror*, September 10, 2022, https://www
        .mirror.co.uk/news/royals/queens-favourite-cars-how-earned-27960067.

2   "When Queen Elizabeth Took King Abdullah for a Drive in Her
        Land Rover," *PBS NewsHour*, January 23, 2015, https://www.pbs.org
        /newshour/world/driving-king-anecdote-goes-viral; see also Sherard

Cowper-Coles, "Secrets of a Diplomat," *Times*, October 14, 2012, https://www.thetimes.co.uk/article/secrets-of-a-diplomat-7kcxbjkksqq.

3   Chris Lezotte, "The Evolution of the 'Chick Car' Or: What Came First, the Chick or the Car?" *Journal of Popular Culture* 45, no. 3 (2012): 516–31, https://onlinelibrary.wiley.com/doi/10.1111/j.1540-5931.2012.00942.x.

4   Katie Sehl, "The Bizarre History of Trying to Design Cars for Women," *VICE*, April 23, 2019, https://www.vice.com/en/article/43jvvb/cars-for-women-design-history-volvo.

5   Alex Mayyasi, "How Subaru Came to Be Seen as Cars for Lesbians," *Atlantic*, June 22, 2016, https://www.theatlantic.com/business/archive/2016/06/how-subarus-came-to-be-seen-as-cars-for-lesbians/488042/.

6   Katherine J. Parkin, *Women at the Wheel: A Century of Buying, Driving, and Fixing Cars* (Philadelphia: University of Pennsylvania Press, 2017), 156.

7   Parkin, *Women at the Wheel*, 155.

8   James B. Twitchell, *Adcult USA: The Triumph of Advertising in American Culture.* (New York: Columbia University Press, 1996), 4.

9   Joseph Serna and Veronica Rocha, "Bruce Jenner Fights Wrongful-Death Lawsuit in Fatal Malibu Crash," *Los Angeles Times*, May 28, 2015, https://www.latimes.com/local/lanow/la-me-ln-bruce-jenner-lawsuit-in-fatal-malibu-crash-20150528-story.html.

10   "Ricky Gervais Slams Caitlyn Jenner in Golden Globes Monologue," YouTube, January 10, 2016, https://www.youtube.com/watch?v=86CT53xvOFY.

11   Caitlyn Jenner's YouTube channel, https://www.youtube.com/watch?v=q-vJ1xtksEM.

12   Manal al-Sharif, *Daring to Drive: A Saudi Woman's Awakening* (New York: Simon & Schuster, 2017).

13   "Row over *Vogue Arabia* Cover Featuring Saudi Princess in Driving Seat," *Guardian*, May 31, 2018, https://www.theguardian.com/world/2018/may/31/row-over-vogue-arabia-cover-featuring-saudi-princess-in-driving-seat.

14   Fatma Tenis, "Saudi Activist Who Led Campaign to Legalize Driving for Women Is Released from Jail," NPR, February 10, 2021, https://www.npr.org/2021/02/10/966258281/loujain-al-hathloul-saudi-activist-jailed-for-driving-has-been-released.

Chapter 12

1   Tom Wolfe, *The Kandy-Kolored Tangerine-Flake Streamline Baby* (New
        York: Picador, 2009), 102.

2   Julie Beck, "The Decline of the Driver's License," *Atlantic*, January 22,
        2016, https://www.theatlantic.com/technology/archive/2016/01/the
        -decline-of-the-drivers-license/425169/.

3   Charles L. Marohn, *Confessions of a Recovering Engineer: The Strong
        Towns Vision for Transportation in the Next American City* (Hoboken,
        NJ: Wiley, 2021), 42.

4   Matthew B. Crawford, *Why We Drive: Toward a Philosophy of the Open
        Road* (Boston: Mariner Books, 2020), 248.

5   "Dashcam Video from inside Driver's Car Shows Him Shooting
        during Road Rage Incident," YouTube, January 28, 2022, https://www
        .youtube.com/watch?v=saQ72NZtrS0.

6   Crawford, *Why We Drive*, 254.

7   Christina Zdanowicz et al., "The Waukesha Victims Included an 8-Year-
        Old Boy, A Loving Grandmother and a Woman Excited to Make Her
        Debut in the Dancing Grannies," CNN, October 26, 2022, https://www
        .cnn.com/2021/11/22/us/waukesha-wisconsin-parade-victims/index.html.

8   Gregory H. Shill, "How Vehicular Intimidation Became the Norm,"
        Atlantic, November 3, 2020, https://www.theatlantic.com/ideas
        /archive/2020/11/how-trump-train-trucks-became-a-political
        -weapon/616979/.

9   "List of Vehicle-Ramming Incidents during George Floyd Protests,"
        Wikipedia, https://en.wikipedia.org/wiki/List_of_vehicle-ramming
        _incidents_during_George_Floyd_protests.

10   "'Trump Train' Ambushes Biden-Harris Campaign Bus in Texas,"
        YouTube, October 31, 2020, https://youtu.be/i0jglSIgRbM.

11   Kate McGee, "911 Transcripts Filed in Updated 'Trump Train'
        Lawsuit Reveal San Marcos Police Refused to Send Escort to Biden
        Bus," *Texas Tribune*, October 29, 2021, https://www.texastribune
        .org/2021/10/29/trump-train-texas-highway-crash-police/.

12   Shill, "How Vehicular Intimidation Became the Norm."

Chapter 13

1   Nancy A. Nichols, "The Car Office Is the Ultimate Sign of Our
        Times," *Boston Globe*, February 17, 2021, https://www.bostonglobe
        .com/2021/02/17/opinion/car-office-is-ultimate-sign-our-times/.

2   "Sometimes we even just say we're going to dinner and instead just park the SUV in the woods, have a couple drinks and get down to biz w/out the worry of being too noisy if you know what I mean." See Alina Dizik, "Parents Pile into Work Conferences to Escape Their Families," *Wall Street Journal*, July 20, 2022, https://www.wsj.com /articles/parents-pile-into-work-conferences-to-escape-their-families -11658323598?mod=hp_featst_pos5.

3   Kate Kellan, "Good Vibrations? COVID Quiet Time Soothes Earth's Seismic Shakes," Reuters, July 23, 2020, https://www.reuters.com /article/us-health-coronavirus-vibrations/good-vibrations-covid-quiet -time-soothes-earths-seismic-shakes-idUSKCN24O2P2.

4   Emma Bubola, "With Sensors on Streets, France Takes Aim at 'Noise from Hell,'" *New York Times*, February 21, 2022, https://www.nytimes .com/2022/02/21/world/europe/france-street-noise.html.

5   "2020 Fatality Data Show Increased Traffic Fatalities during Pandemic," NHTSA, June 3, 2012, https://www.nhtsa.gov/press-releases/2020 -fatality-data-show-increased-traffic-fatalities-during-pandemic.

6   Eugenia Akhim, "The Ugly Truth behind the Fastest Cannonball Run Ever," HotCars, June 10, 2022, https://www.hotcars.com /fastest-cannonball-run/.

7   Andrew Brinker, "'Really Concerning Behavior': Continuing Pandemic Trend, Traffic Deaths Rose Again in 2021," *Boston Globe*, January 23, 2022, https://www.bostonglobe.com/2022/01/23/metro /really-concerning-behavior-continuing-pandemic-trend-traffic-deaths -rose-again-2021/?p1=BGSearch_Overlay_Results.

8   David Leonhardt, "Race, Class and Traffic Deaths," *New York Times*, August 23, 2022, https://www.nytimes.com/2022/08/23/briefing /traffic-deaths-class-race-covid.html.

9   Emily Badger and Alicia Parlapiano, "The Exceptionally American Problem of Rising Roadway Deaths," *New York Times*, November 27, 2022, https ://www.nytimes.com/2022/11/27/upshot/road-deaths-pedestrians-cyclists .html#:~:text=In%202021%2C%20nearly%2043%2C000%20people,—%20 cyclists%2C%20motorcyclists%2C%20pedestrians.

10   Badger and Parlapiano, *New York Times*, November 27, 2022.

11   Mike Colias and Nora Eckert, "A New Brand of Sticker Shock Hits the Car Market," *Wall Street Journal*, February 26, 2022, https://www .wsj.com/articles/a-new-brand-of-sticker-shock-hits-the-car-market -11645851648?mod=hp_lead_pos11.

12   Arlie Hochschild and Anne Machung, *The Second Shift: Working Families and the Revolution at Home* (New York: Penguin Books, 2012).

13   "Why Women Need the Office," *Economist*, August 28, 2021, https://www.economist.com/business/2021/08/28/why-women -need-the-office.

Chapter 14

1   Caroline Criado Perez, *Invisible Women: Data Bias in a World Designed for Men* (New York: Abrams, 2019), 186.

2   Joel W. Eastman, *Styling vs. Safety: The American Automobile Industry and the Development of Automotive Safety, 1900–1966* (Lanham, MD: University Press of America, 1984), 7.

3   Matt Gurwell, "Meet Sierra Sam—The Original Crash Dummy," *Legal Examiner*, March 23, 2022, https://affiliates.legalexaminer.com /transportation/sierra-sam/.

4   "NHTSA's Crash Test Dummies," NHTSA, https://www.nhtsa.gov /nhtsas-crash-test-dummies.

5   Keith Barry, "The Crash Test Bias: How Male-Focused Testing Puts Female Drivers at Risk," *Consumer Reports*, October 23, 2019, https ://www.consumerreports.org/car-safety/crash-test-bias-how-male -focused-testing-puts-female-drivers-at-risk/.

6   "Lawmakers Push DOT to Update Female Crash Test Dummies," *Safety+Health*, May 2, 2022, https://www.safetyandhealthmagazine .com/articles/22541-lawmakers-push-dot-to-update-female-crash -test-dummies.

7   Susan Molinari and Beth Brooke, "Women Are More Likely to Die or Be Injured in Car Crashes: There's a Simple Reason Why," *Washington Post*, December 12, 2021, https://www.washingtonpost .com/opinions/2021/12/21/female-crash-test-dummies-nhtsa/.

8   Barry, "Crash Test Bias."

9   Barry, "Crash Test Bias."

10   "Female Crash Fatality Risk Relative to Males for Similar Physical Impacts," NHTSA, August 2022, https://crashstats.nhtsa.dot.gov /Api/Public/ViewPublication/813358.

11   Emily Featherston, "Federal Agency Touts 'Accelerated Efforts' to Implement More-Advanced Crash Test Dummies, Promises Continued Research into Safety Disparities," *InVestigate*, October 21,

2022, https://www.investigatetv.com/2022/10/21/federal-agency
-touts-accelerated-efforts-implement-more-advanced-crash-test
-dummies-promises-continued-research-into-safety-disparities/.

12   Maria Weston Kuhn, "It's Time to End Discrimination in Crash
Testing," *Ms.*, August 10, 2022, https://msmagazine.com/2022/08/10
/discrimination-women-crash-test-dummies-car/.

13   Kuhn, "It's Time to End Discrimination in Crash Testing."

14   David Leonhardt, "Race, Class and Traffic Deaths," *New York Times*,
August 23, 2022.

15   Kristoffer Tigue, "Heavy Electric Vehicles Jeopardize Climate Action
and Public Safety, Experts Warn," Inside Climate News, February 7,
2023, https://insideclimatenews.org/todaysclimate/heavy-electric
-vehicles-jeopardize-climate-action-and-public-safety-experts-warn/.

16   Maurie Backman, "New Data Reveals That Car Ownership Costs
Women Up to $7,800 More than Men," Motley Fool, March 26,
2022, https://www.nasdaq.com/articles/new-data-reveals-that-car
-ownership-costs-women-up-to-%247800-more-than-men.

17   Steve Tengler, "New 'Pink Tax' Study Shows Women Pay Upwards
of $7,800 More for Car Ownership," *Forbes*, October 27, 2021, https
://www.forbes.com/sites/stevetengler/2021/10/27/new-pink-tax-study
-shows-women-pay-upwards-of-7800-more-for-car-ownership/?sh
=52b6272463d.

18   Andrew Ross and Julie Livingston, "Once You See the Truth About
Cars, You Can't Unsee It," *New York Times*, December 15, 2022.

19   Gagosz, Alexa. "How a Hard-Working, Middle-Class Family
Spiraled Into Homelessness," *Boston Globe*, December 17, 2022.

20   Barchie, Get Back to Normal, Go Fund Me, https://www.gofundme
.com/f/a-family-get-back-to-normal.

Chapter 15

1   Warren James Belasco, *Americans on the Road: From Autocamp to Motel,
1910–1945* (Cambridge, MA: MIT Press, 1979).

2   "Electric Vehicles," IEA, https://www.iea.org/reports/electric-vehicles.

3   "California Moves to Accelerate to 100% New Zero-Emission Vehicle
Sales by 2035," California Air Resources Board, August 25, 2022,
https://ww2.arb.ca.gov/news/california-moves-accelerate-100-new
-zero-emission-vehicle-sales-2035.

4   Dan Neil, "2023 Cadillac Lyriq: An Electric Leap Forward," *Wall Street Journal*, December 1, 2022, https://www.wsj.com/articles /2023-cadillac-lyriq-an-electric-leap-forward-11669927707?page=1.

5   Caleb Miller, "Dodge Charger Daytona EV Concept Back for SEMA, and an Even Wilder Banshee Is Coming," *Car and Driver*, October 31, 2022, https://www.caranddriver.com/news/a41823742 /dodge-charger-daytona-ev-concept-sema-power-details/.

6   Elon Musk, "The Mission of Tesla," November 18, 2013. https://www .tesla.com/blog/mission-tesla.

7   "At the heart of this mobility experience is the word 'feel,'" Mizuno said, explaining that focus will be on sensing and interacting with people. See Peter Valdes-Dapena, "Sony and Honda Reveal Their New Car Brand," CNN, January 5, 2023, https://www.cnn.com /2023/01/04/tech/sony-honda-afeela/index.html.

8   Nathan Bomey, "Why Tesla 'Has a Problem Appealing to Women': Electric Cars, Elon Musk May Be Off-Putting," *USA Today*, July 8, 2019, https://www.usatoday.com/story/money/cars/2019/07/08/tesla -electric-cars-women-elon-musk/1616632001/

9   Jennifer Hiller, "Tesla to Open Some of Its Charging Network to Other EVs, White House Says," *Wall Street Journal*, February 15, 2023, https ://www.wsj.com/articles/tesla-to-open-some-superchargers-to-other -vehicles-white-house-says-2ed35fd5?mod=hp_lista_pos4.

10   Aaron Pressman, "Massachusetts Needs More EV Chargers—Wicked Badly," *Boston Globe*, August 11, 2023, https://www.bostonglobe.com /2023/08/11/business/massachusetts-needs-wicked-more-ev-chargers /#:~:text=Massachusetts%20will%20need%20to%20dramatically ,according%20to%20a%20new%20report.

11   Jack Ewing and Peter Eavis, "Electric Vehicles Start to Enter the Car-Buying Mainstream," *New York Times*, November 13, 2022, https://www.nytimes.com/2022/11/13/business/electric-vehicles -buyers-mainstream.html.

12   Sabrina Shankman and Aaron Pressman, "We Drove around New England Looking for EV Chargers: It Was Better than Expected," *Boston Globe*, January 10, 2023, https://www.bostonglobe.com/2023 /01/10/science/we-drove-around-new-england-looking-ev-chargers -best-doughnuts-it-was-easier-than-we-expected/.

13   Aaron Pressman, "Mass Pike EV Chargers Took the Holiday Weekend Off," *Boston Globe*, May 30, 2023, https://www.bostonglobe

.com/2023/05/30/business/mass-pike-ev-chargers-took-holiday
-weekend-off/#:~:text=The%20electric%20vehicle%20chargers%20
installed,t%20totally%20out%20of%20luck.

14   Dan Neil, "Thank Tesla: The Biggest Obstacle to EV Ownership
      Will Soon be History, *Wall Street Journal*, July 20, 2023, https://www
      .wsj.com/articles/thank-tesla-the-biggest-obstacle-to-ev-ownership
      -will-soon-be-history-dc9945be).

15   Jack Ewing, "Tesla Will Open Some Chargers to All Electric
      Vehicles," *New York Times*, February 15, 2023.

16   "Batteries for Electric Vehicles," US Department of Energy Alternative Fuels
      Data Center, https://afdc.energy.gov/vehicles/electric_batteries.html.

17   Dan Neil, "Can the New Chevy Bolt EUV Win Back America's
      Trust?" *Wall Street Journal*, August 2, 2022.

18   Hiller, "Tesla to Open Some of Its Charging Network to Other EVs."

19   Chevy advertisement, *Wall Street Journal*, January 3, 2023, B5.

20   It is important to note here that this was not the case for many women
      of color, as they were often denied access to Whites-only bathrooms.
      For more history, see Susan V. Spellman, "All the Comforts of Home:
      The Domestication of the Service Station History, 1920–1940," *Journal
      of Popular Culture* 37, no. 3 (2004): 463–77, https://onlinelibrary.wiley
      .com/doi/10.1111/j.0022-3840.2004.00080.x; Nathaniel Meyersohn,
      "Believe It or Not, Gas Stations Used to Be Squeaky Clean: Here's
      What Changed," CNN Business, June 4, 2022, https://www.cnn
      .com/2022/06/04/business/gas-station-bathrooms-history.

21   Phoebe Wall Howard, "She is Chief Engineer of All-Electric Ford
      F-150, Leading a Revolution," *Detroit Free Press*, May 17, 2021.

22   A lone holdout seems to be Toyota, a company that is hedging their
      bets with other things such as in hydrogen. See River Davis and Sean
      McLain, "Toyota Chief Says 'Silent Majority' Has Doubts about
      Pursuing Only EVs," *Wall Street Journal*, December 18, 2022, https
      ://www.wsj.com/articles/toyota-president-says-silent-majority-has
      -doubts-about-pursuing-only-evs-11671372223?mod=hp_lead_pos5.

23   Tom McCarthy, *Auto Mania: Cars, Consumers, and the Environment*
      (New Haven, CT: Yale University Press, 2007), 253.

24   Leo Marx, *The Machine in The Garden* (Kiribati: Oxford University
      Press, 2000), 7.

25   Austin Lott, "2022 GMC Hummer EV Weight, Range and Battery
      Size Revealed," Edmunds, February 17, 2022. https://www.edmunds

.com/car-news/2022-gmc-hummer-ev-range-battery-pack-and
-weight-revealed.html

26   Vanessa Friedman, "When the Czarinas Ruled the Front Row," *New York Times*, July 4, 2022, https://www.nytimes.com/2022/07/04/style /mira-duma-elena-perminova-ukraine.html.

27   Widely attributed to Alfred Einstein, but likely a paraphrase from his *Collected Writings*. For more see *The Ultimate Quotable Einstein* (Princeton, NJ: Princeton University Press), 2010.

Chapter 16

1   Lauren Smiley, "'I'm the Operator': The Aftermath of a Self-Driving Tragedy," *Wired*, March 8, 2022, https://www.wired.com/story /uber-self-driving-car-fatal-crash/.

2   Lauren Smiley, "The Legal Saga of Uber's Fatal Self-Driving Car Crash Is Over," *Wired*, March 28, 2023.

3   Kirsten Korosec, "Rivian Hit with Gender Discrimination Lawsuit That Alleges Toxic 'Bro Culture,'" *TechCrunch*, November 4, 2021, https://techcrunch.com/2021/11/04/rivian-hit-with-gender -discrimination-lawsuit-that-alleges-toxic-bro-culture/.

4   Laura Schwab, "Life outside the Boys Club: Why I Spoke up about Rivan's Toxic Bro Culture (and Got Fired)," *Medium*, November 4, 2021, https://medium.com/@lauraschwab1/life-outside-the-boys -club-why-i-spoke-up-about-rivians-toxic-bro-culture-and-got-fired -15600b999ae7.

5   Victoria Scott, "Your Car Is Tracking You Just as Much as Your Smartphone Is—and Your Data Is at Risk," Drive, June 8, 2022, https://www.thedrive.com/news/think-your-smartphone-is -tracking-you-your-cars-doing-it-too.

6   "Vehicle-to-Everything," Wikipedia, https://en.wikipedia.org/wiki /Vehicle-to-everything.

7   Chieko Tsuneoka, The Coming U.S.-China Race Over 5G in Cars, *Wall Street Journal*, May 27, 2021.

8   Scott, "Your Car Is Tracking You Just as Much as Your Smartphone Is."

9   "'We Are No Longer the Customers': Inside the 'Age of Surveillance Capitalism,'" WBUR, January 15, 2019, https://www.wbur.org /onpoint/2019/01/15/surveillance-capitalism-online-data-privacy -facebook-google-amazon.

10   Adam Satariano, "Meta's Ad Practices Ruled Illegal under E.U. Law," *New York Times*, January 4, 2023, https://www.nytimes.com/2023/01/04/technology/meta-facebook-eu-gdpr.html.

11   Matt Egan, "Credit Card and Car Loan Delinquencies Pass Pre-Covid Levels as Consumers Get Squeezed," CNN, August 10, 2023, https://www.cnn.com/2023/08/10/economy/credit-card-car-loan-pay-failure-pre-covid/index.html.

12   Shoshana Zuboff, *The Age of Surveillance Capitalism: The Fight for a Human Future at the New Frontier of Power* (New York: PublicAffairs, 2019), 191.

13   Scott, "Your Car Is Tracking You Just as Much as Your Smartphone Is."

14   Matthew Crawford, *Why We Drive: Toward a Philosophy of the Open Road* (New York: Morrow, 2020), 41.

15   "Mom Approved: Two Busy Partners Share Why They Think Autonomous Driving Could Help Families and Small Businesses," Waymo, December 2, 2022, https://ltad.com/news/az-mom-squad-approved.html.

Chapter 17

1   Dana Goldstein et al., "Suburban Women, No Longer 'Soccer Moms,' Hold Key to Midterms," *New York Times*, November 4, 2022, https://www.nytimes.com/2022/11/04/us/suburban-women-midterms.html.

2   That scene was inspired by the real-life experience of Academy Award–winning screenwriter Callie Khouri. Khouri was walking home from her shift as a waitress when a man in a parked car bold-facedly asked her to "suck his dick." It is hard to overstate how common this was for Baby Boomer women, who had the good fortune to have some opportunities in the larger world but also faced decades of harassment on the street and on the job.

3   Crawford, *Why We Drive*, 314.

4   Many thanks to Rochelle Roberts, who drafted and researched this report on *American Auto* for this section of the book.

5   Jeff Peek, "Pony Cars Stamps Will Make It Cool to Write Letters," *Hagerty*, July 18, 2022, https://www.hagerty.com/media/news/pony-cars-stamps-make-it-cool-to-write-letters/.

6   Charles L. Marohn, *Confessions of a Recovering Engineer: The Strong Towns Vision for Transportation in the Next American City* (Hoboken, NJ: Wiley, 2021), 20.

7   Crawford, *Why We Drive*, 217.

Index